MICRO HYDROELECTRIC POWER STATIONS

MICRO HYDROELECTRIC POWER STATIONS

By

L. MONITION, M. LE NIR and J. ROUX

translated by
Joan McMullan

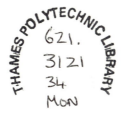

A Wiley–Interscience Publication

JOHN WILEY & SONS
Chichester · New York · Brisbane · Toronto · Singapore

First published as *Les microcentrales
hydroélectriques* © 1981 Masson SA, Paris.

Copyright © 1984 by John Wiley & Sons Ltd.

All rights reserved.

British Library Cataloguing in Publication Data:

Monition, L.
   Micro hydroelectric power stations.
   1. Water-power electric plants
   I. Title   II. Le Nir, M.   III. Roux, J.
   IV. Les microcentrales hydroelectriques.
   English
   621.31'2134   TK1081

ISBN 0 471 90255 1

Library of Congress Cataloging in Publication Data:

Monition, Lucien.
   Micro hydroelectric power stations.

   Translation of: Les microcentrales hydroélectriques.
   'A Wiley–Interscience publication.
   Includes index.
   1. Hydroelectric power plants—Design and construction.
I. Le Nir, M.   II. Roux, J.   III. Title.
TK1081.M5513 1984   621.31'2134   84–7454

ISBN 0 471 90255 1 (U.S.)

Typeset by Photo·Graphics, Honiton, Devon.
Printed by St Edmundsbury Press, Bury St Edmunds, Suffolk.

Contents

Foreword

L'Agence Française pour la Maîtrise d'Énergie (AFME) is very pleased that John Wiley & Sons Ltd have decided to translate and publish *Les microcentrales hydroélectriques* (which was orginally published in 1981 by Masson SA), as once again it illustrates the interest in micro hydroelectric power stations for the decentralized production of electrical energy and their potential in both the industrialized and the developing countries.

In France, there have been a number of consequences resulting from the legislation of 15 July 1980 on space heating and energy savings. Permission was given to local authorities to produce electricity using micro hydro-power stations and to sell it to the national grid, which holds the monopoly on electricity transmission and distribution. Local authorities were given the power to authorize installations of up to 4500 kW per site, subject to the requirement for a detailed evaluation of the environmental impact.

These new conditions have involved an increase in the number of requests to build micro installations, particularly from the local authorities themselves. This development has aroused anxiety in other users of the watercourses, whether for irrigation, water distribution, angling, or canoeing, and particularly among ecologists who regard micro hydro-power stations as a threat to the balance of nature and an attack upon the countryside.

AFME has participated in general diagnostic studies and has given help with feasibility studies in order to encourage the development of projects which allow for such technical and environmental objections.

It is important that the installations should be sized to suit the available hydrological resources, and, to this end, a data bank is being created which will hold information on harnessable energy in France. Two national bodies (the Office for Geological and Mining Research (BRGM) and the Centre for Agricultural Mechanization of the Department of Waterways and Forests (CEMAGREF)) are collaborating in this project following the AFME's initiative.

Micro hydro-power will not solve the energy crisis, but well designed installations can satisfy many demands under acceptable financial and environmental conditions. In particular, they allow independent action at the local level.

In this book L. Monition, M. Le Nir, and J. Roux consider most of the problems posed by the creation of micro hydro-power stations either in the developing countries or in industrialized countries.

The wide range of topics covered includes evaluation of the resource and the conditions for its viability, implementation of the civil engineering works under conditions suited to local economic constraints, environmental impact and appropriate countermeasures, and the selection of the electromechanical equipment to suit various discharges and heads.

Finally, this very informative book concludes with an overview of the situation for small hydro-power stations throughout the world, and it will certainly make a significant contribution to the implementation of well balanced projects.

Michel Rolant,
President,
I 'Agence Française pour la Maîtrise de l'Énergie,
Paris, France

Preface to the English edition

The harnessing of the energy of running water to provide mechanical power has been one of man's great engineering achievements, and yet unlike the other great innovations, such as the steam engine or the internal combustion engine, it is not associated with any particular inventor. Its development cannot be traced to a definite time, though references begin to occur in Asia Minor in the 1st century BC. Even by the beginning of the 18th century, the power required for grinding and water pumping could be provided by the Roman and Norse wheels that were still in common use. The industrial revolution created new requirements, however, which demanded larger and larger power-generating capacities, and led to rapid improvements in the designs of water wheels and turbines. The capacities of water-power plants became very large indeed as electricity generation became commonplace near the end of the 19th century, and the position was ultimately reached whereby large-scale hydroelectric plants now contribute anything from 2% to 80% of total national electricity demand. The possibility of increasing this proportion is very limited in most industrialized countries as most potential large-scale sites have already been exploited.

The oil price shocks of 1973 and 1979 have created an interest in reversing the trends again. It is now more expensive to generate and distribute electricity from large thermal power stations, with the result that small communities are beginning to turn again to the possibility of providing at least part of their electrical requirements from renewable resources. Small hydroelectric power stations can make an important contribution in this area, and they have an equally important role to play in the underdeveloped countries where the cost of diesel fuel has made it impossible to provide power supplies to rural communities. Indeed, there is even a case for enlightened self-interest on the part of the industrialized countries as the markets for exporting both equipment and expertise are expanding rapidly throughout the world.

Unfortunately, every improvement has its price and the price for exploiting water power is vigilance in ensuring that the environment is not irreparably damaged and that life in the rivers can continue to flourish. This requires considerable care and attention to detail.

In this book, the authors discuss all aspects of hydro-power exploitation using micro hydro-power stations, which are small systems suitable for single dwellings or small communities. They introduce the problems of assessing the power available from a given stream and of designing and building the siteworks. They discuss the various types of turbine, and the conditions under which they can be employed. They describe the environmental hazards that are posed by micro hydro installations and the measures that should be adopted to counteract them, and they discuss the economics of micro hydroelectricity projects. All of this is important and useful, but in many ways their other topic of concern is even more important. A detailed discussion is included of the legislative procedures that are followed in France to actively encourage the development of small-scale water power while ensuring that the interests of all other water users are also protected. The French procedure for obtaining permission to build a micro hydro-power station includes an investigation of the uses to which the power is to be put and an examination of the implications that the proposed installation will have on regional planning. This demonstrates a welcome awareness of the larger issues involved and is an approach that could be pursued with benefit in many other countries.

J. T. McMullan
Editor,
International Journal of
Energy Research.

CHAPTER 1

General considerations

'Hydropower is a completely national resource, which can be exploited without importing any other fuel. ... The harnessing of water power offers an insurance against external factors. ... It is infinitely renewable and immune to fluctuations in the economy. One further general point is that diversification of energy sources is in itself a safety factor' (Report by Sénateur Pintat, 1976).

'... Expenditure on equipment for harnessing small waterfalls should be encouraged by simplifying and speeding up bureaucratic formalities, and by offering the same financial inducements as are given for energy conservation...' (Report adopted by the Conseil Economique et Social sur les Perspectives Energétiques, *Journal Officiel*, 28 June 1979).

'... Consequently, everything possible should be done to facilitate the maximum exploitation of natural energy resources, such as hydro-power. In rural areas, there is often a shortage of power for the development of small-scale industries. This has repercussions on job creation...' (Statement by the Groupe Artisanat, *Journal Officiel*, 28 June 1979).

The energy crisis produced by the increased price of oil has led to a worldwide renewal of interest in all exploitable energy resources. Hydroelectricity is one area which has seen a great resurgence of interest, to such an extent that all countries with hydraulic resources (including those where large-scale sites have already been exploited) are now seriously considering the development of low power hydroelectricity.

The industrialized countries are again begining to install this type of equipment after revising their ideas on its economic feasibility as, over the last two decades, it has not been economic in comparison with fossil fuels. As a result, the production of suitable sturdy and reliable plant to standardized designs is also evolving rapidly.

The developing countries, which have been very seriously affected by the oil price rises (and which also have technical difficulties in maintaining diesel generators), are encouraging the increased use of hydroelectric power through developing a better knowledge of their available water resources.

Hydraulic energy is a form of non-polluting solar energy which can be easily converted into electricity with an efficiency of up to 70%. However, it does pose environmental problems, and, no matter how important the energy

1

contribution, the harnessing of low power hydroelectricity should not be undertaken without fully considering the other calls upon the water source, including leisure activities.

1.1 ATTEMPT AT A DEFINITION

The term 'micro hydroelectric power station' (MHPS) defines an installation for the production of hydroelectricity at low power levels. In practice, the power from such installations can be between 5 and 5000 kW for heads of 1.5–400 m and flows ranging from several hundreds of litres per second to several tens of cubic metres per second.

In most cases, the power stations are sited directly on a watercourse, with no control reservoir, and they require a good knowledge of the stream's flow pattern in order to size the turbines correctly to control the production. The civil engineering works related to the water intakes are often carried out at a minimum cost using locally available materials.

A simple classification can be drawn up in terms of the head and the hydromechanical equipment involved:

Power (kW)	Head (m)		
	Low	Medium	High
5–50	1.5–15	15–50	50–150
50–500	2–20	20–100	100–250
500–5000	3–30	30–120	120–400

Several attempts at classification have been made in different countries and one common definition is to use *micro* when the power is less than 100 kW, and *mini* when it is between 100 and 5000 kW.

1.2 POWER AND ENERGY

The hydraulic power which is naturally available at a given site is defined by

$$P = g\,\rho QH,$$

where P is the hydraulic power (in watts), g is the acceleration due to gravity (equal to 9.81 m s^{-2}), ρ is the liquid density (in kilograms per cubic metre), Q is the flow or discharge (in cubic metres per second), and H is the height of the waterfall or head (in metres).

In the case of water ($\rho = 1000$ kg m^{-3}), this equation becomes

$$P = 9.81 \; QH.$$

The corresponding electrical production can then be written as

$$W = Pt \; \eta \; f,$$

where W is the electrical energy produced (in kilowatt-hours), P is the hydraulic power (in kilowatts), t is the operating time (in hours), equal to $24 \times 365 = 8760$ h yr^{-1}, η is the efficiency of the turbine–generator assembly, usually between 0.5 and 0.9, and f is a coefficient to allow for seasonal flow variations for run-of-river installations.

Summary 1   The various elements of a micro hydroelectric power station

Intake

This involves a structure whose shape and dimensions are determined by local engineering conditions. The intake can be flooded and the dams can be constructed of concrete, rockfill, gabions (wire baskets filled with stones), earth, or bricks made with local materials.

Head-race or full-pipe

The head-race can be made of earth or concrete. If earth is used, the velocity of the water should not exceed 0.80 m s^{-1}, but with concrete it can reach 1.5 m s^{-1}.

For electricity generation the intake can be in an irrigation channel or on a drinking water supply.

Full-pipes are usually made of steel or polythene tubing and allow water speeds of 3–8 m s^{-1}.

Intake grill and settling pond

The turbine must be protected from drifting material carried by the river. This can be done using a grill which is kept clean either manually or by means of a simply designed scraper. A settling pond also allows the sedimentation of small particles before the water enters the main system.

Sluice-gates

Several types of gate are used depending on whether they are to protect against a rise in the river water-level, to isolate the channel or to isolate the turbine.

Turbines

The choice of equipment depends on whether the installation is to be connected to the national grid or is to provide a self-sufficient electricity supply. If the former, then it will be sized for maximum electricity production, while in the latter case continuity of production will be the prime consideration.

The type of turbine selected depends on the head of water involved. Propeller or Kaplan turbines are used for low heads. They consist of an axial shaft to which are attached adjustable blades which can be trimmed during operation. Francis turbines, in which the water is directed on to a vaned waterwheel, are used for medium heads. Pelton turbines, in which, a water-jet, under very high pressure, is directed by means of an injector on to buckets on the circumference of the wheel are suitable for high heads. The use of the injector permits accurate control of the flow of water to the wheel. Banki–Mitchell (or Ossberger cross-flow turbines) can be designed to suit a large range of heads and water flow rates.

Regulator

The regulator ensures that a constant rate of rotation is maintained regardless of the network demand. It can be mechanically or electrically operated. In the mechanical type, the admission of water to the turbine is controlled. As demand increases, the turbine tends to rotate more slowly, and the regulator increases the supply of water to the turbine which then returns to its normal speed. When demand decreases, the operation is reversed. In the electrical type, passive resistances are used to absorb the difference in energy.

Alternators

The turbine is connected to an alternator either directly or through a gearbox. Asynchronous units are used for grid-connected installations and synchronous units are used for isolated installations.

Transport and distribution networks

The energy is produced as a low tension three-phase (220–380 V) supply and can be used on site for local consumption. If the site where the electricity is to be used is far from the source (over 1 km away) it will usually be preferable to transform it from the low tension ($<$ 500 V) to a medium tension ($\sim$ 20 000 V). This will require the installation of a step-up transformer and one or more step-down transformers at the end of the medium tension line. The cost of a medium tension line is about 90 000 FF per kilometre.

Safety regulations specify conditions on private installations, whether domestic or industrial, to avoid short circuits and accidental earthing.

1.3   HYDROELECTRICITY AND MICRO HYDROELECTRIC POWER STATIONS

The natural hydroelectric resources of the world have been estimated (using 1974 data) at 36 000 TWh per year ($9.81QH \times 8760$). The hydraulic reserve which represents the technically and economically exploitable fraction of the total resources is about 9800 TWh. The geographical distribution is as follows (in terawatt-hours): Europe (excluding the USSR), 700; USSR, 1100; USA, Canada, and Greenland, 1300; Japan and China, 1450; Central and Latin America, 1850; Africa, 2000; Asia (excluding Japan, China, and Siberia), 1200; Oceania, 200.

In 1978 the total world production was 1560 TWh, or 15% of the exploitable potentials. While in some countries the degree of exploitation has been large – 65% in Europe excluding the USSR, 64% in Japan, and 40% in

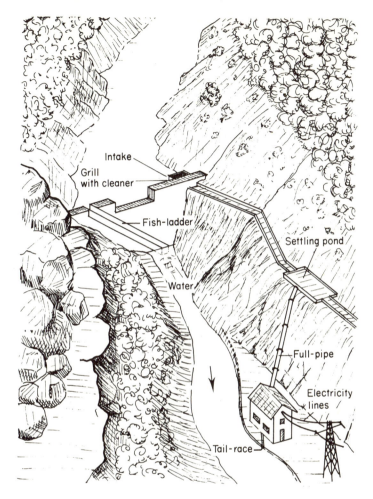

Fig. 1 Schematic diagram of a micro hydroelectric power station

the USA and Canada – this is not always the case, particularly among the developing countries – 7% in Central and Latin American, 1.7% in Africa, and 6.2% in Asia excluding Japan, China, and Siberia.

In France, the national utility, Éléctricité de France (EDF), has estimated the total potential at 270 TWh and the harnessable potential at 80 TWh in a survey which mainly involved sites with a power output greater than 2000 kW. Of the harnessable potential as defined in this way, large installations ($P \geqslant$ 2000 kW) provide 66–70 TWh, and small installations ($P \leqslant$ 2000 kW) provide 5–10 TWh. When those installations already in operation are excluded, some 9–18 TWh are still available.

In 1977 France had 1060 micro hydroelectric power stations (MHPSs) of less than 2000 kW. This corresponds to a total of 390 MW and a production of

2 TWh which represents 3% of total hydro-electric production. Of this, 1.6 TWh is produced by 'autonomous', or independent producers and 0.4 TWh is produced by EDF. (2 TWh is equivalent to 450 000 toe, or 3 300 000 barrels of oil.)

In 1979, by including MHPSs of powers up to 5000 kW, a figure of 1170 MHPS is reached (217 of which are operated by EDF). This represents a power of 0.75 GW for a production of 3.5 TWh.

The entire production outside EDF up to a threshold of 8000 kVA is 4.7 TWh, which represents 7% of hydroelectric production or 2% of total French electricity production. It yields an annual savings of 1 mtoe (7 350 000 barrels of oil).

Very few countries have attempted a precise evaluation of their low power hydroelectric potential. In the light of the conditions imposed by the oil crisis, it would seem advisable that such surveys should be undertaken.

Summary 2   Role of hydroelectricity in the production and consumption of energy in France (in mtoe)

		Coal	Oil	Natural Gas	Nuclear	Hydro	New energy sources	Total	Percentage hydroelectric (%)
1960	Produced	38.9	2.2	2.9		9		53	17
	Consumed	46.8	26.9	2.9		9		85.6	11
1965	Produced	36.4	3.7	5	0.2	10.3		55.6	19
	Consumed	45.7	49.7	5.1	0.2	10.5		111.2	9
1973	Produced	19.4	2	7	2.7	10.6	2	43.7	24.7
	Consumed	30.5	117	15	3.1	9.8	2	177.4	6
1978	Produced	15.9	1.9	7.1	6.4	15	3	49.3	30
	Consumed	32.4	108.8	20.8	6.5	16	3	187.5	9
1979	Produced	15.4	3	7	8.4	14.5	3	50.3	29
	Consumed	34.6	108.5	23.2	8.4	15.8	3	193.5	8
1980	Produced	15.2	2.2	7	15.2	15.3	3.2	58.1	26.3
	Consumed	34.4	101.7	23.6	12.9	16	3.2	191.7	8.3
1981*	Produced	15.6	2.4	6.6	22.1	15.8	3.4	65.9	24
	Consumed	32.1	90.7	24.7	22.1	14.7	3.4	187.7	7.8

* Provisional figures.
Source: Les chiffres clés de l'énergie, 1982–3.

1.4 UTILIZATION, COSTS AND REGULATIONS

The use of low power hydro-energy in the developed countries, while it is totally dependent upon the availability of water, i.e. upon the level of the river, can meet the requirements of local industries or isolated villages. In France, since 1980, such communities have been allowed to produce energy for community needs in this way and to derive income by selling their surplus to EDF.

In the developing countries, low power hydroelectricity can light villages and pump water for both people and livestock, and for irrigation, and can provide power for the conservation of foodstuffs, medicines, etc.

The cost of the system varies from one installation to another and could be reduced by standardization of equipment. As a general rule, the cost decreases as the unit power increases, and, for comparable powers, low head installations are more expensive. Civil engineering costs are an important factor for any installation and can represent between 40 and 80% of the total capital cost, depending upon local geological and technical conditions. In France, the price of an installation, including the civil engineering works, can vary from 5000 to 15 000 FF per kilowatt.

In some countries where construction conditions are very difficult and where the technology has to be imported, e.g. Burundi or Zaire, these costs can be tripled. They can be halved, however, if the civil engineering works are carried out using local materials, as is done in Indonesia, Peru, and China.

Regulations on the use of rivers for the production of energy differ from country to country. In France, EDF has a monopoly on the transmission and distribution of electricity and on production at levels above 8000 kVA (or about 7000 kW, allowing for the power factor). For installations of over 4500 kW, a licence must be applied for, and a permit from the relevant local authority is sufficient for powers below this level.

1.5 ENVIRONMENTAL IMPACT

If micro hydroelectric power stations are not well managed, they can cause damage to the natural environment. The removal of water can create difficulties for migrating fish and so there should be a bypass between the intake and the outlet with a stipulated minimum reserved flow. At certain times of the year, the upstream reservoir can be liable to eutrophication, with all the biological effects that this involves.

In France, only a 'statement' of the impact is required for MHPSs with powers of less than 500 kW, but a more detailed 'study' is required for powers above this level. In both cases, however, the evaluation of the effects upon the environment must be accompanied by concrete suggestions for repairing the damaging effects by the building of fish-passes, control of the discharge from the reservoir, and maintenance of water levels.

1.6   INSTALLATION CRITERIA

The seven principal criteria involved in the installation of a micro hydroelectric power station are as follows:

(a) Knowledge of the various water users throughout the drainage basin (both present and future needs), the water rights already granted, and the minimum required flows (e.g. for the breeding of fish).

(b) Knowledge of local requirements, with an examination of the present energy supply. The proposed applications of the energy produced by the micro power station must be evaluated and a selection must be made.

(c) Determination of harnessable heads.

(d) Knowledge of the hydroclimatology of the area, i.e. the pattern of rainfall and the discharges of the rivers. If necessary, when actual measurements are not available, rainfall and temperature data can be analysed to determine the details of the flow.

(e) Knowledge of the hydrogeology of the area, noting the role of aquifers in the supply and flow of the rivers, particularly during low flow periods.

(f) Knowledge of the geology of the site and determination of those areas which are liable to movement of the soil and subsoil. This is required in order to define the stability of the various structures so that the civil engineering works can be carried out under the best technical and financial conditions.

(g) Analysis of the environmental impact, particularly as it affects soil drainage upstream of the water intake, and disturbance caused to fish breeding. Definition of the associated site works that are required.

1.7   USEFUL DATA AND CONVERSION FACTORS

Power

Power is the amount of energy (work or quantity of heat) supplied per second. The *watt* (W) is the unit of power:

$$1 \text{ watt (w)} = 1 \text{ joule per second } (\text{J s}^{-1})$$

$$1 \text{ kilowatt (kW)} = 10^3 \text{ W}$$

$$1 \text{ megawatt (MW)} = 10^6 \text{ W} = 10^3 \text{ kW}$$

$$1 \text{ gigawatt (GW)} = 10^9 \text{ W} = 10^3 \text{ kW}$$

$$1 \text{ terawatt (TW)} = 10^{12} \text{ W} = 10^9 \text{ kW}$$

$$1 \text{ horsepower (hp)} = 0.735 \text{ kW}$$

(*Note*: The horsepower is no longer a recommended unit.)

Total electric power installed in France	45 000 MW
Hydroelectric power	17 000 MW

Installed power at Genissiat	400 MW
Light water nuclear power station	
at Fessenheim or Bugey	900 MW

Energy

The unit for work or the consumption of energy is the *joule*.

1 joule $= 0.2778 \times 10^{-6}$ kWh

$\qquad = 0.239 \times 10^{-3}$ millitherms (mth) or kilocalories (kcal)

kWh $= 1$ kW $\times$ 3600 s $=$ 3600 kJ $=$ 860 kcal

1 kilowatt-hour $\qquad = 10^3$ Wh

1 megawatt-hour (MW h)$= 10^3$ kWh

1 gigawatt-hour (GW h) $= 10^6$ kWh

1 terawatt-hour (TWh) $\quad = 10^9$ kWh

	1977	1979
Total electrical production in France	202 TWh	231 TWh
Hydroelectricity contribution	76 TWh	66 TWh
Produced by EDF	71 TWh	60 TWh
Produced by independent producers	5 TWh	6 TWh

Production by a nuclear power station of 900 MW	5 TWhyr^{-1}

Energy Equivalents

1 tonne of oil equivalent (toe)
$\qquad = 10 000$ therms
$\qquad = 1.5$ tonne of coal equivalent
$\qquad = 11 625$ kWh of natural gas
$\qquad = 4500$ kWh of electricity
$\qquad = 11 625$ kWh of heat

1 tonne of oil	= 1 toe
1 tonne of coal	= 0.67 toe
10 000 kWh of natural gas	= 0.86 toe
10 000 kWh of electricity	= 2.22 toe
10 000 kWh of heat	= 0.86 toe

Power and average annual consumption of some domestic appliances

Washing machine	2.8–3.5 kW	450 kWh
Dish washer	3 kW	1050 kWh
Refrigerator	0.2 kW	290 kWh
Freezer	0.3 kW	800 kWh
Television set	0.2–0.4 kW	180 kWh
Vacuum cleaner	0.3 kW	360 kWh
Iron	0.8 kWh	150 kWh
Electric cooker	6 kW	1000 kWh
Lighting	0.05–1 kW	300 kWh
Water heating		1800 kWh
Central heating	12–15 kW	20 000–30 000 kWh

Temperature conditions (France) during heating season

Minimum outside temperature	$-7\,°C$
Maximum outside temperature	$+15\,°C$
Average outside temperature	$+4\,°C$
Length of heating season	October–April

EDF has suggested various classifications of supply depending upon the amount of electrical equipment in a household and the conditions under which it is used. These were

Basic household requirements	3 kW
Comfortably equipped	6 kW
Well equipped	9 kW

For a house of 100 m² with floor area with all-electric central heating, the minimum power required is 12 kW, but some domestic supplies exceed 36 kW.

Some energy requirements (in toe and kWh)

Construction of a detached house, 100 m² area	10 toe	45 000 kWh
Heating	2–3 toe yr^{-1}	9000–12 500 $kWh\,yr^{-1}$
Lighting	0.2 toe yr^{-1}	900 $kWh\,yr^{-1}$
Cooking for four people	0.2 toe yr^{-1}	900 $kWh\,yr^{-1}$
Vehicle, 1 tonne	1.3 toe	5850 kWh
Manufacture of 1 tonne of		
Synthetic fabric	5.9 toe	26 550 KWh
Aluminium	5.1 toe	22 950 kWh
Recycled aluminium	0.18 toe	810 kWh
Polystyrene	3.7 toe	16 650 kWh
Asbestos cement	1.4 toe	6300 kWh
Glass	0.6 toe	2700 kWh
Steel	0.6 toe	2700 kWh
Paper	0.5 toe	2250 kWh
Cement	0.11 toe	495 kWh
Brick	0.07 toe	315 kWh

(*Source*: EDF)

BIBLIOGRAPHY AND FURTHER READING

ANONYMOUS (1978). Bilan énergétique français en 1978. *Pétrole informations*, no. 1499, 40–43.

ANONYMOUS (1978). Utilisation de l'énergie hydraulique. *Le Pont*, February 1978, special issue.

BRIN, A. and MONITION, L. (1979). Microcentrales hydroélectriques. *Entretiens écologiques de Dijon, Cahiers trimestriels*, no. 3/4, 26–29.

COTILLON, J. (1979). Micro-power: an old idea for a new problem. *Water Power and Dam Construction*, **31** (1).

DIETHRICH, R. (1979). Les minicentrales hydroélectriques. *Le Particulier* (to appear).

MAUCOR, J. P. (1980). *Les microcentrales hydrauliques*, Ministère de la Coopération, Paris.

MINISTÈRE DE L'INDUSTRIE ET DE LA RECHERCHE (1976). *La production d'électricité d'origine hydraulique*, Les dossiers de l'énergie, no. 9, La Documentation française, Paris.

MINISTÈRE DE L'INDUSTRIE (1980). *Les chiffres clés de l'énergie, 1979*, La Documentation française, Paris.

MONITION, L. (1979). Preliminary study of OLADE's regional programme on small hydroelectric plants for Latin America. Contribution as official French government expert to a UNIDO seminar workshop, Katmandu, 10–14 September 1979.

MONITION, L. (1980). *Les microcentrales hydroélectriques. Comité consultatif pour la Recherche et le Développement de l'Énergie*, Paris, La Documentation française.

UNITED NATIONS (1977). *Les methodes utilisées pour evaluer les services rendus par les centrales hydrauliques. Rapport français*, Ep/GE3/R30/Add 1, United Nations, Geneva.

VARIOUS AUTHORS (1979). *Microcentrales hydroélectriques*, special issue no. 4 of *Annales des Mines*, Paris.

CHAPTER 2

Hydraulic resources

2.1 WATER AS AN ENERGY RESOURCE: TEMPORAL AND SPATIAL VARIATIONS

Hydraulic potential is the combination of the possible flows and the distribution of gradients, and the hydraulic resource is that fraction of the hydraulic potential which is still accessible after allowing for economic considerations. A micro hydroelectric power station can divert only a part of this, the exploitable resource, and it thus recovers the potential energy of the water which would have been dissipated to no benefit in the natural flow along the watercourse.

Hydro-power owes its position as a renewable resource to the variable but more or less continuous flow of a certain amount of water in the stream. This water, supplied by the rain and always moving, constantly flows from the continents to the sea, where it evaporates back into the atmosphere in an unending cycle controlled by two opposing forces, the heat of the Sun and the Earth's gravity.

Any water deposit, such as a glacier, a river or a subterranean body of water, is therefore a dynamic object which fluctuates under the influence of *gains* from rain, *losses* through evaporation or consumption, and flow reductions caused by obstacles. These include turbulence in a torrent or soil permeability leading to the entrapment of water which is continually trying to reach its lowest gravity potential, the sea.

The various time-lags introduced while traversing this geological maze modulate the initial variability of the rains and determine the different patterns of the streams which form an image of the rain which has fallen on the catchment area after filtering through a complex geological and topographical environment (Fig. 2).

The existence of the geological 'filter' causes the flow of the watercourse at a given moment to depend upon earlier climatic events. The delay in the response of the catchment area is a function of the various delays in the arrival of the components of the flow at the measuring station. These time-lags can range from several hours to several months. The flow of a watercourse cannot therefore be divorced from the notion of duration and

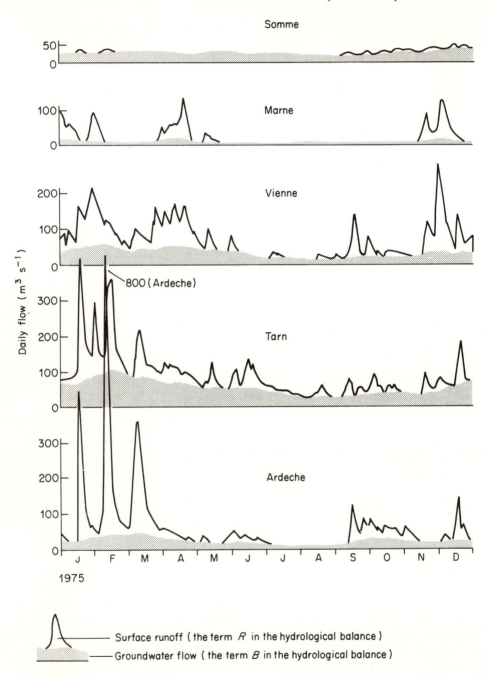

Fig. 2 Examples of the effect of geological and topographical filters on the conversion of rainfull into water-flow, for various regions of France (from Bodelle and Margat, 1980)

should be understood not as a unique static value but as a population of values each having its own probability of existence over a certain period of time, whether a year, a month, or a few days.

It is therefore more convenient to speak of the flow of a stream in terms of *characteristic flows*, which are abstract concepts chosen as a function of the desired application. These characteristic flows will be determined by *statistical analysis* of measurements of the instantaneous flow at different parts of the catchment.

2.2   WATER AS AN ENERGY RESOURCE: CLIMATIC VARIATIONS

2.2.1   The water cycle and the drawing up of a balance sheet – the variable nature of the resource

Over a sufficiently large area, say some tens of square kilometres, and over a sufficient time period, it can be assumed that the equilibrium equation

$$P = R + I + E \tag{1}$$

is valid, where P is the precipitation, R is the runoff, I is the infiltration, and E is the evapotranspiration. This relation describes an average equilibrium where each term is continually being modified by the forces of gravity and the energy of the Sun (Fig. 3).

This continual readjustment is represented by the dynamics of the *water cycle*, which can be divided arbitrarily into four phases:

(a) Condensation of the water vapour in an air mass to produce *precipitation (P)*.
(b) Evaporation of water directly by contact with the air or indirectly through the consumption or transpiration of vegetation. This is *evapotranspiration (E)*.
(c) *Infiltration (I)* of water into subterranean nappes. Some of this water rejoins the watercourse through drainage to form the *base flow (B)* at low water.
(d) *Runoff (R)* of non-infiltrated water, resulting from drainage of the soil. This runoff combines to form part of the flow of the river and to provide drainage to the subterranean nappes.

The discharge or flow (Q) of a stream which is useful for the production of hydroelectricity is the sum of the runoff (R) and of the drainage (B). This can be expressed as

$$Q = B + R = P - E - I + B \tag{2}$$

The runoff and infiltration are often combined under the term *effective rain* $(R_e = R + I)$ which represents the amount of water available for both the

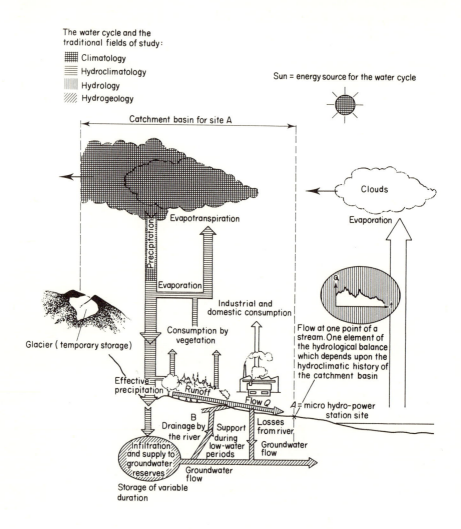

Fig. 3 Schematic water cycle for a catchment basin:

$$Q = B + R = P - E - I + B$$

Note: The relative value of the terms in the hydrological balance depends upon the surface areas, the climate, and the geology of the region; national balances have been established over long periods:

France	Evapotranspiration $\simeq 61\%$ of precipitation (Margat, 1980)
USA	Evapotranspiration $\simeq 75\%$ of precipitation (Remenieras, 1972)
West Germany	Evapotranspiration $\simeq 52\%$ of precipitation (Keller and Claudius, 1955)

infiltration and the runoff, since there is a constant interchange between these two terms at each point of the catchment area.

The variation in the flow is determined by the variations in the supply (the atmosphere), in the evapotranspiration rate (the temperature), in the infiltration (the degree of saturation of the soil), and in the drainage (the position of the nappe in relation to the river).

Precipitation, the starting point of the cycle, varies greatly with time. The best meteorological forecasts do not exceed 24 h and, even then, they only predict the average behaviour over a region, i.e. over an area of several hundred or thousand square kilometres. Further, the forecast is qualitative by nature and does not indicate the amount of expected precipitation over the given surface area.

It must be taken, therefore, that knowledge of precipitation levels is subject to uncertainty, which can only be allowed for by using a long-term statistical study with data collected over a period of 30 years. It has been demonstrated in this way, for example, that there is a relationship between the value of the mean annual precipitation and its annual distribution.

The flow of a stream, since it results from the precipitation on the local catchment area, thus exhibits the same random character, and similar statistical techniques will be required to determine its characteristics. Nevertheless, there are cases where the flow at a particular point of a river has not been measured over a sufficiently long period to allow a direct study and the extended series of data forming a flow history has to be reconstructed using the terms in equation (2).

Before describing the methods for determining the useful flow (which is related to the end use and the prevailing economic conditions), it is therefore convenient to examine the various terms of equation (2) and the factors to which they relate.

2.2.2 The hydrological balance: measurements and calculation

(a) *Precipitation (P)*

Precipitation is the amount of atmospheric water which falls on a region in terms of its projected horizontal area, either in liquid form as rain or in solid form as snow or hail. It is measured at different rain measuring stations and is represented by the isohyets on rainfall maps.

Measurement of precipitation Precipitation is measured by the amount of water collected in a rain or snow gauge, and is represented by the depth of the water layer which would have accumulated over a given horizontal surface if all of the precipitation had been captured, i.e. by the volume of water collected divided by the cross-sectional area of the rain gauge. Although simple, this definition conceals a number of difficulties which are a source of error and lead to inaccuracy in the measurement. For example, whatever type of rain gauge is used, it produces an aerodynamic disturbance which affects the amount of rain collected. Thus the positioning, the size and the height of the rain gauge all influence the measurements, and it is advisable to use standard apparatus in order to be able to compare measurements with each other. A rain gauge can be of the recording type or of a simple type which does not record the measurements. The choice of rain gauge depends upon the purpose of the measurement, the capacity of storage required, the type of precipitation, and the desired frequency of the observations. The gauge installed thus represents a compromise between technical requirements and the available budget. France has on average one rain gauge per 150 km^2, Italy has one per 80 km^2, and England has one per 40 km^2. A further factor is that there should be more stations in mountainous regions than on plains to allow for the greater variation in precipitation.

Typically, measurements are performed by various bodies such as the National Meteorology Office, the electricity utilities, the Highways Department, the Forestry Commission, the Army, Navy, etc., and by volunteers. The measurements are usually collated centrally, and they are generally published in a number of summaries giving statistics with varying degrees of detail.

Data analysis and calculation of the amount of water which has fallen on a catchment area   The number of years over which observations have been made affects both the desired characteristic value and its accuracy. Accuracy increases with the number of years for which records are available, but the acquisition of data will be expensive. It is advisable, therefore, to know the minimum number of years required to achieve a sufficiently accurate evaluation. Binnie has studied the effect of length of time of data collection on the value of the annual rainfall average for 53 stations distributed over different countries (Table 1). It can be seen that a minimum of 10 years of measure-

Table 1   Effect of number of years measurement on the accuracy of the average (after Binnie)

Number of years	Variation of the percentage error relative to the calculated value over a long period, with the number of years in the period
1	+51 to −40%
2	+35 to −31%
3	+27 to −25%
5	+15 to −15%
10	+ 8 to − 8%
20	+ 3 to − 3%
30	+ 2 to − 2%

ments is needed in order to give an accuracy of better than 10%. With more than 20 or 30 years of records, the value calculated varies only by a few per cent.

Statistical analysis of the measurements is necessary because of the large volume of data to be processed and the random nature of the precipitations. The analysis allows the isolation of parameters or characteristic values in terms of the particular application so as to summarize the measurements in the minimum possible number of values which can be used for representing them in a calculation, or on a map or graph.

The data are first analysed to produce a central or dominant value such as

(i) The daily, monthly, seasonal or annual mean
(ii) The inter-monthly average (the average for the same month taken over several years),

$$\bar{x} = \sum_{i=1}^{n} \frac{x_i}{n}$$

(iii) The rainfall index, which is the ratio of the annual (or monthly) precipitation to the average annual (or monthly) precipitation
(iv) The rainfall coefficient, which is the ratio of the actual precipitation over a period to the average annual precipitation over the same period.

This dominant value is then related to the other observations using the dispersion parameters:

－   the variation, $\omega = x_{max} - x_{min}$
－   the frequency distribution curve
－   the average absolute deviation, $e_{aver} = \dfrac{1}{n} \sum_{i=1}^{n} [x_i - \bar{x}]$

－   the standard deviation, $\sigma = \sqrt{\dfrac{(x_i - \bar{x})^2}{n - 1}}$.

The basic data required for the hydrological balance is the precipitation on the catchment area. Various methods have been developed to determine this from the measurements recorded at different rain-measuring stations within the catchment area, and the polygon, or Thiessen, method is often the most suitable (Fig. 4).

(b)   *Evapotranspiration (E)*

Evapotranspiration is the sum of two terms:

(i) *Direct* evaporation from areas of open water (rivers, lakes, glaciers, etc.) and the wet areas of the catchment basin (soil, vegetation, buildings, roads, etc.). This is a physical phenomenon involving the transformation of liquid water into water vapour.

(ii) *Indirect* evaporation provided by the transpiration of the plants. This is a biological phenomenon.

This is an arbitrary distinction and depends upon how one regards the water cycle. Natural consumption by the vegetation (which excludes that involved in irrigation) is included here under indirect evaporation and not as a deduction from a previously determined resource.

In fact, evaporation from open surfaces, soil and plant cover, and the transpiration of plants, are difficult to measure separately without carrying out many observations. These processes have thus been combined under the general term *evapotranspiration*.

This combination is justified in part because many of the parameters (such as temperature, humidity, insolation, and pressure) are common to the fundamental terms. Simple expressions have therefore been sought to allow direct calculation of the evapotranspiration (E). As this is related to the amount of water available, the calculation methods first require quantification of the maximum possible evapotranspiration by assuming that the amount of water available is unlimited. This potential evapotranspiration (E_p) is then compared with the actual availability of water for the selected period to find

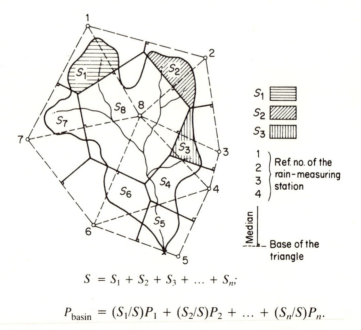

$$S = S_1 + S_2 + S_3 + \ldots + S_n;$$

$$P_{\text{basin}} = (S_1/S)P_1 + (S_2/S)P_2 + \ldots + (S_n/S)P_n.$$

Fig. 4 Calculation using the polygon method of the amount of water falling on a catchment area of surface area S containing n measuring stations

the real evapotranspiration (E_r) and the term (E) in the equation for the hydrological balance. With most modern methods of calculating the real evapotranspiration, the availability of water in the soil is treated as a quantity of water (the useful reserve, UR) present in a soil of maximum capacity, URMAX. If the precipitation is insufficient to supply the evapotranspiration (maximum evapotranspiration = potential evapotranspiration), then the evapotranspiration requirement is met by recovering part or all of the reserve.

Summary 3 Examples of equations for calculating the average potential evapotranspiration during a particular month

---

Penmann's equation

$$E = 0.22 \times 10^3 \, (q_s - q) \, (0.93 + u_2),$$

where E is the evaporation rate from shallow open water in kilograms per square metre per day, q_s is the saturation mass concentration of water vapour in the air (humidity) at ambient temperature, q is the actual mass concentration of water vapour measured above the evaporation surface, and u_2 is the wind speed (in metres per second) measured at 2 m above the evaporating surface.

This equation has yielded good results over a large range of climates, with a slight tendency to overestimate the evaporation, although it does require a knowledge of parameters which are rarely available.

Turc's equation

$$ET_p = K(I_g + 50) \, t(t + 15),$$

where ET_p is the potential evapotranspiration expressed in millimetres per month, t is the average monthly temperature in degrees Celsius for the month considered, and I_g is the average global solar radiation, in calories per square centimetre per day.

$$I_g = I_{gA} \, (0.18 + 0.62 \, h/H),$$

where I_{gA} is the radiant energy which would reach the Earth's surface in the absence of the atmosphere, in calories per square centimetre per day, h is the period of effective insolation in hours, H is the length of the day in hours. K is a coefficient which is dependent on the calculation period and the relative humidity (r.h.) of the soil.
  (a) *Turc's period equation*: $K = 0.13$ for periods of 10 days; $K = 0.123$ for periods of 8 days; $K = 0.143$ for periods of 11 days.
  (b) *Turc's monthly equation*: $K = ab$, where $b = 1$ if r.h. $> 50\%$ for every month of the year, $b = 1 + (50 - \text{r.h.})/70$ if r.h. $< 50\%$ for one or more months of the year, $a = 0.37$ for February and $a = 0.4$ for the other months. In practice, for our latitudes, r.h. $> 50\%$ for every month of the year.

Thornthwaite's equation

$$ET_p = 1.6 \, (10 \, t/I)^\alpha,$$

where ET_p is the monthly evapotranspiration in millimetres for a fictitious month of 30 days and a theoretical insolation period of 12 h per day, t is the average monthly temperature in degrees Celsius for the month under consideration, and

$$\alpha = 0.016I + 0.5,$$

where I is the sum of the monthly indices i throughout the year, and $i = (t/5)^{1.514}$.

This equation can be criticized because of the large number of numerical coefficients of dubious accuracy. Furthermore, it involves the value of an annual index I determined by extrapolation of the value of the index i into the future.

L. Serra's equation

$$ET_p = 22.5\,[(1 - \epsilon_m)/0.25]\,[1 - (\tau^2/1000)]\,e^{0.0644\,T_a}$$

where T_a is the average temperature for the month in degrees Celsius, τ is the half-amplitude of the extreme variations in the monthly temperatures, ϵ_m is the monthly wet bulb temperature of the air, and ET_p is the average monthly evapotranspiration in millimetres.

If the precipitation is sufficient, it will first satisfy the requirement for evapotranspiration (E) and then complete the useful reserve (UR) of the soil. If there is still an excess, this forms the effective rain (R_e), the amount of water available for runoff and infiltration.

This calculation is made for a determined time period, such as a day, a week or a month. After each calculation step, the value of the useful reserve of the soil has changed, and the calculation can be performed again for the next time period. While this method is a simplification of the natural phenomenon, it does allow the variations in the surface availability of the resource to be estimated.

The flow chart of Fig. 5 shows the stages of the calculation. These can be performed either manually or on a computer.

An example of calculation by the manual method is given in Table 2.

(c) *Infiltration (I) and drainage of underground water by streams (B)*

The effective rain as determined in the preceding section represents the availability of water for *runoff* and *infiltration* into the ground water.

The hydrological balance established for France by J. Margat (1980) shows that the transfer of ground water to the sea is less than 1% of the precipitation while infiltration is 23% of precipitation. Thus, for large hydrographic basins, discontinuities in the geological strata mean that almost all of the drainage of the ground water (B) is by surface flows.

For the outlet of such basins, and of smaller basins which can be regarded as being impermeable, the measured flow therefore represents the precipitation reduced solely by the evapotranspiration, because in the equation of balance (equation (2)) either

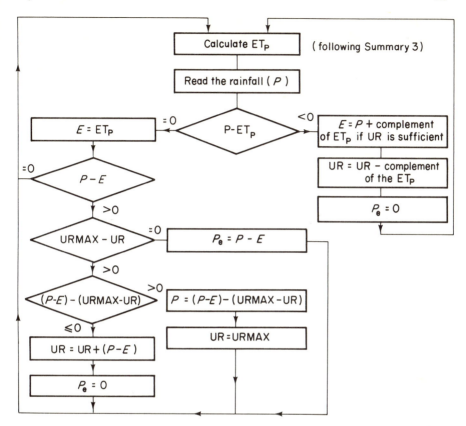

Fig. 5 Schematic flow diagram of the calculation of the evapotranspiration (E) and the effective precipitation (P_e). ET_p is the potential evapotranspiration; E is the actual transpiration; P is the precipitation; P_e is the effective precipitation (water available for runoff and infiltration); UR is the water available in the ground store for evapotranspiration; URMAX is the maximum capacity of the ground store. All these variables are measured in millimetres per calculation period (day, 10-day period or month)

$$I = B \quad \text{(large basins)} \quad \text{and} \quad Q = P - E$$

or

$$\begin{cases} I &= 0 \quad \text{(small impermeable basins)} \\ B &= 0 \end{cases} \quad \text{and} \quad Q = P - E$$

In spite of this equivalence in the balance equation, the hydrological behaviour of the two types of basin are fundamentally different. For large basins, the term B represents water which has remained underground

Table 2 Example of manual calculation using the flow chart for real evapotranspiration (E): calculation of cumulative infiltration and of evapotranspiration in time steps of 5 days in the particular case of a semi-arid zone

	Calculation period																							
	November 1975			December 1975					January 1976						February 1976				March 1976					
Precipitation, P (mm)	0	0	23	0	0	90	0	32	11	9	9	50	45	0	62	12	0	0	23	0	75	0	44	10
Potential evapotranspiration, ET_p (mm)	34	34	34	30	30	30	30	30	31	31	31	31	31	34	25	25	25	25	15	15	15	15	15	18
$P - ET_p$ (mm)	−34	−34	−11	−30	−30	+60	−30	+2	−20	−22	−22	+19	+14	−34	+37	−13	−25	−25	+8	−15	+60	−15	+29	−8
UR*† (mm)	0	0	0	0	0	+40	+10	+12	0	0	0	+19	+33	0	+37	+24	0	0	+8	0	+40	+25	+40	+32
I*‡ (mm)	0	0	0	0	0	20	20	20	20	20	20	20	20	20	20	20	20	20	20	20	40	40	54	54
E (mm)	0	0	23	0	0	30	30	30	23	9	9	31	31	33	25	25	24	0	15	8	15	15	15	18

* URMAX, the maximum useful reserve of the soil is 40 mm.

† UR is the useful reserve of the soil.

‡ I is the total infiltration into the underground nappe. In this example, $I = R_c$ because there is no runoff, i.e. $R = 0$.

(sometimes for a long time) before becoming a component of the surface flow again, whereas for a small impermeable basin there is no term *B* which would represent a delayed contribution and the basin responds to the precipitation more quickly and more abruptly.

In contrast, the return of ground water to the surface system is spatially progressive, and, for small catchment basins of interest for micro hydroelectric stations, the infiltrated water only *partially* returns to the watercourse. The remainder forms the groundwater flow in Fig. 3, and equation (2) balances only if this new term is included.

Direct measurements of the infiltration use essentially agronomical techniques. It is possible to use point measurements of good accuracy to evaluate the infiltration for an area as small as a cultivated field (which has homogeneous characteristics), but it is very difficult to extrapolate such measurements to the whole catchment area, where vegetation cover, the gradient, and the type of soil change from place to place.

A detailed determination of the contribution of ground water to surface streams involves hydrogeological techniques, but for micro hydroelectric stations the exploitable flow is more commonly estimated by indirect methods of determining the global infiltration (*I*) and the drainage (*B*).

Nevertheless, as these methods approximate those which allow the flow a river to be calculated from the precipitation, they will be discussed in the next section.

2.3 THE FLOW OF A WATERCOURSE: CERTAINTIES AND UNCERTAINTIES IN THE RESOURCE

Measurement of the flows and analysis of the behaviour of a watercourse is a necessary prologue for any study of the hydraulic management of a basin.

The equation for the hydrological balance, as defined earlier, theoretically permits the calculation of the flow of a stream using climatological measurements. However, uncertainties in the values, either because of systematic errors due to the measurement method or because of errors inherent in the simplifying hypotheses adopted, result in a preference for direct processing of measurements taken at the station nearest to the site being studied. This is possible if recording has been carried out for a period of greater than ten years.

Occasionally, available data cannot be used, e.g. where there are karstic soils resulting in a lack of proportionality between the flows and the catchment area on account of the dominance of losses or reappearances in the watercourse. These can cause sudden changes in flows by either abruptly removing water or introducing water which originated elsewhere.

In the remainder of this section, methods will be given for the recording and presentation of results and the official recording stations will be described, including their distribution both in siting and administration terms. Methods for using the results will then be given, both for a long series of records applying to a neighbouring station and for a short series of data recorded over a period of at least one year. In the latter case, the flow measurements are related to a long series of results available for a neighbouring station or to representative climatological data.

Because of the large volume of data to be handled, these methods usually involve computer processing but, for each method, a manual approximation technique will be developed which is often sufficient at the preliminary study stage.

2.3.1 Measurement of flow in a watercourse

The methods available fall into two principal groups: *mechanical methods*, which involve a calibrated weir or direct measurement of the flow velocity through a known cross-section, and *chemical methods*, where the dilution of a product added at a known concentration is measured.

(a) *Weirs*

This type of station is the most frequently used. The depths of the water are either measured once or twice per day with a staff gauge, or they are recorded continuously with an automatic recorder.

The conversion from water depth to flow is performed using a calibration curve established previously either by direct measurements of the flow or by theoretical calculation relating the flow through an opening of known geometrical shape to the depth of water upstream of the weir (and eventually to the depth of water downstream in the drowned-flow case).

Gauging the amount of water in small streams (with flows of less than $5 \text{ m}^3 \text{ s}^{-1}$ can be carried out with a thin plate weir. A weir is an orifice with an open top. It is usually rectangular in cross-section and has vertical sides (see Fig. 6).
be created for measuring specific flow ranges.

Summary 4 gives the equations used for the main types of weir. These are simple to use, and lend themselves well to the evaluation of the flow in a particular application. Accuracies of 1–2% can be achieved, provided that certain operating precautions are taken.

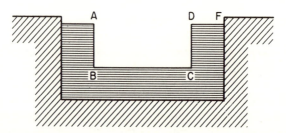

Fig. 6 Rectangular weir: CD and AB are the flanks; BC is the crest; DF is the side-contraction

Summary 4 Principal types of gauging weirs and their ranges of application:

$Q = \mu\Omega \sqrt{2gh}$, where Ω is the flooded cross-section and μ is the flow coefficient

Rectangular vertical thin plate weir

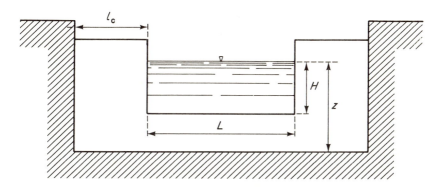

$$Q = \mu LH \sqrt{2gH},$$

where L is the width of the sill; $(z - H)$ is the crest height of the weir.
    H should be measured upstream of the weir at a distance of at least $4H$ in order to avoid the influence of the drop in water-level at the sill.
    l_c is the side-contraction

(a) If $l_c = 0$ (i.e. there is no side-contraction), then

$$\mu = (0.405 + (0.003/H)) [1 + 0.55 (H/(H + z))^2]$$

and for $0.5 < L < 2$ m, $0.1 < H < 0.6$ m, and $0.2 < z < 2$ m we have $\mu \simeq 0.43$.
(b) If $l_c \neq 0$ (i.e. there *is* side-contraction), then we have

$$\mu = (0.405 - 0.003 (l_c/(L + l_c) + 0.0027/H) [1 + 0.55(LH/((l_c+L) (H+z)))^2]$$

   (Hégly's equation). With $0.1 < H < 0.6$ m, $0.4 < L < 1.8$ m, $0.4 < z < 0.8$ m, and $0 < l_c/(L + l_c) < 0.9$, we find that generally $\mu \simeq 0.40$.

Triangular thin plate weir

(This type of weir maintains good accuracy even with small flows.)

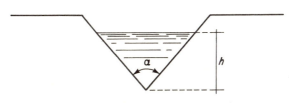

$$Q = 1.32 \tan (\alpha/2) h^{2.47}.$$

This is known as the Gourley and Grimp equation; Q is the flow in cubic metres per second and h is in metres.

$$\text{For } \alpha = 90°, \, Q = 1.32 \, h^{2.47}$$

$$\text{For } \alpha = 60°, \, Q = 0.76 \, h^{2.47}$$

$$\text{For } \alpha = 45°, \, Q = 0.55 \, h^{2.47}$$

(*Note*: $h^{2.47}$ can be calculated directly using most small pocket calculators.)

It is preferable to use a weir with side-contraction in order to have a free nappe. This type of weir is cut from a sheet and has dimensions appropriate to the flow to be measured (Fig. 7). Preferably, $H = 0.05$ m and $h \geqslant 0.05$ m (Fig. 8).

An arrangement of planks can also be used with a metal plate installed on the upstream side of the opening, above the sill and the flanks, in order to form the thin plate which is needed for the equations given above.

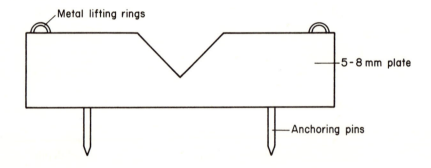

Fig. 7   Sheet metal weir for measurements on small streams

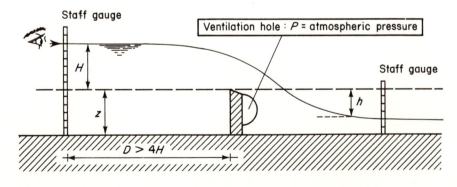

Fig. 8   Measurement of the depth of water in the discharge

When the plate cannot be regarded as being thin, the general equation for the weir, $Q = \mu\Omega \sqrt{2gH}$, is still valid, but the coefficient μ must be measured by calibration at the station, or on a scale model. This type of weir is not suitable, therefore, when the measurement programme requires a quick, inexpensive estimate of the potential of a low power site.

Table 3   Choice of weir for measuring small flows ($Q \leqslant 5$ m^3 s^{-1})

Range of flows to be measured (l s^{-1})		Type of weir to be used	Corresponding gauged head (cm)		Comments
Minimum	Maximum		Minimum	Maximum	
1	140	Triangular, $\alpha = 90°$	5	40	If possible the crest height should be >0.45 m so that the channel depth is >0.85 m
20	200	Rectangular, 0.30 m wide	10	40	Crest height $\geqslant 0.40$ m
50	500	Rectangular. 1 m wide	10	40	Channel depth $\geqslant 0.80$ m
100	1000	Rectangular. 1–2 m wide	10 to 15	45 to 65	{ Channel depth $\geqslant 0.90$ m { Channel depth $\geqslant 1.30$ m
500	5000	Rectangular. 5–10 m wide	10 to 15	40 to 65	{ Channel depth $\geqslant 0.90$ m { Channel depth $\geqslant 1.30$ cm

H can be measured with a level and scale or with a graduated staff at a distance of at least $4H$ upstream of the weir. The measurement scale is placed on the side of the channel or preferably in a side shaft communicating with the flow through a pipe which has a small cross-section in order to eliminate oscillations.

(b)   *Natural or artificial sills of any shape*

Any shape of natural or artificial sill can be used, such as the floor between the piles of a bridge. In such cases, the equation $Q = f(H)$ is obtained by

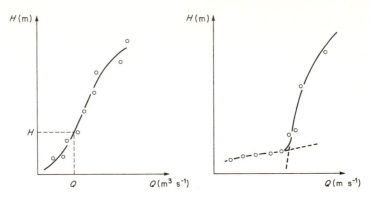

Fig. 9   Examples of calibration curves

calibration of the site for several flow rates ranging from the minimum low flows to the maximum flood flows (Fig. 9).

Water depths are measured by staff gauges set on a mark which is generally situated by one bank of the stream. The staff gauge should be levelled to a fixed reference to allow it to be replaced in case of damage during a flood. They are usually marked at 1 cm intervals, and are generally composed of standard metric segments which can be placed end to end. Devices for the automatic recording of depth measurements are also used.

Flow measurements are made either by using turbine-meters or dilution methods.

Gauging using a turbine-meter   A current meter comprises a propeller whose rotation rate is proportional to the flow rate of the current. The propeller is fixed to a profiled ballast, which allows it to be lowered to the required depth with minimum turbulence. The propeller assembly is lowered to various depths at several points across the river, reached either by means of a metal boom along which it slides (for small flows), or by using a cable which can be moved on a jib forming a cable-car type of structure across the river (for large flows).

The various points on the cross-section of the river can also be explored using a turbine-meter connected to an inflatable dinghy.

The accuracy of measurement using a turbine-meter is almost always better than 3% because systematic errors are few and small, and random errors are reduced because of the large number of measurements made for each section. This good accuracy is obtained when the flow is uniform with few eddies. However, when the flow is not stable (as in a torrent, for example), the streamlines are not parallel because of the turbulence, and turbine-meters are not suitable.

Gauging by chemical dilution   A concentrated solution of a salt is added to a section of the river and its dilution is measured downstream at a sufficient distance to ensure good mixing:

$$Q = A \times \frac{C_{solution}}{C_{downstream}},$$

where A is a coefficient depending upon the material and procedure adopted.

The flow which is measured is the average of the flow at the injection section and that at the measurement section. This value will therefore be closer to the instantaneous flow of the river when the distance between these two is short. The distance which can be used depends on the rate of mixing and the obtaining of a homogeneous dilution. It is therefore more appropriate to use dilution measurements for turbulent flow. There are two principal methods:

(a) By integration, which involves the injection of a certain volume, V, of concentrated solution upstream and repeated downstream sampling over the time, t, between the introduction of the tracer and its disappearance:

$$Q = \frac{V}{t} \times \frac{C_{solution}}{C_{downstream}}$$

(b) By injection at a constant rate and measurement of the development of the concentration downstream until stabilization is reached:

$$Q = q \times \frac{C_{solution}}{C_{downstream}}$$

where q is the injection rate.

The second method has the advantage of being more reliable and more accurate than the first, especially if the experimenter is inexperienced. On the other hand, two to three times as much salt is required, as well as heavier equipment and a longer measurement time.

Sodium dichromate (1), rhodamine B (2), sodium nitrite (3), manganese sulphate (4), sodium chloride (5), lithium chloride (6), and radio-isotopes are all commonly used tracers, as shown in the following table:

	$NaCr_2O_2$ (1)	$C_{10}H_{21}Cl\ O_3N_2$ (2)	$NaNO_2$ (3)	$MnSO_4$ (4)	$NaCl$ (5)	$LiCl$ (6)
Normal solubility $(g\ l^{-1})$	600	10	750	500	200	500
Minimum concentration for direct analysis* $(mg\ l^{-1})$	0.2	0.1	1	2	5–30	0.2

* Concentrations of certain tracers can also be employed prior to analysis, allowing dilutions of some 10–100 times weaker for the injected solution.

The ideal tracer should be very soluble in water, should have good chemical stability when in solution in the stream, should not be adsorbed by the clays in suspension in cloudy water, should not already be present in the river, should be simple to analyse in the weakest possible concentrations, should be cheap, and, most importantly, should be non-toxic to humans and to the flaura and fauna in the concentrations used.

In the past, the most commonly used tracer was sodium dichromate, but the toxicity of the hexavalent chromium ion means that it is no longer as acceptable.

In order to achieve good mixing, the minimum length of the measurement section depends upon the value of the flow and especially upon the width of the river. Thus, to a first approximation, it should be 50–75 times the width of the river. Analysis of the samples is generally performed using a colorimeter, though, for NaCl, by comparing the samples with a series of standard solutions resistivity measurements are used. The accuracy of measurement is 1–2%.

2.3.2   The French gauging network: organization and types of station

The measurement of river flows in France has really only developed since 1900, with the formation of the national hydrology and hydraulic departments.

In comparison with the situation for rainfall measurements, there are very few stations for which more than a century of records are available. One such station is at Rheinfelden on the Rhine near Basle, Switzerland.

Further, the density of the hydrographic network fluctuates and depends on the objectives or financial restrictions of the moment. Thus, there were approximately 1100 hydrometric stations in 1970, and, since then, the emphasis has lain on major power stations either of national interest (first-order stations on the main rivers) or of regional interest (second-order stations). Stations of local interest (third-order stations) and which were sited in small catchment areas allowed the potential of different regions to be evaluated in the past, without reference to the existing energy demand. Unfortunately, the number of point measurements also tends to be lower for such small stations. This can have unfortunate long-term effects, as, ideally, the measurement network should be sufficiently complex to provide good data for actual needs some 10 or 20 years after measurements begin. It is quite difficult to foresee new requirements over such an extended period.

The methods of measurement presently used require extensive manpower for the manual reading of water depths, the production of depth records and measurements of flow using turbine-meters or chemical dilution methods. However, these techniques are being superseded by automatic data collection methods which can take readings over longer time intervals and can transmit the data directly to the recording centre by radio or telephone.

In France many organizations are involved with water-flow measurement. Some examples are as follows:

(a) The Ministry of Agriculture, whose stations are controlled by the agriculture and water services departments. These are often only of a regional or local nature (third-order stations) and concern small or medium-sized catchment areas of non-privately owned rivers. They are usually the most recent stations, having been in operation for less than ten years, and the data they provide, therefore, has to be extrapolated or expanded to cover a longer period before statistical processing can be performed for application to a micro hydroelectric power station.

(b) The Ministry of Transport, which controls its stations through its Supplies Department. The principal aim of these stations is to forecast flooding, and so they are well equipped to measure variations in water depth.

(c) The electricity utility (EDF), whose stations are intended to allow the company to manage the dams on its hydroelectric schemes.

(d) The Ministry of the Environment, whose central hydrological services and related finance departments operate gauging stations, frequently for a particular project.

(e) The Ministry for Industry, which uses its Inter-regional Industrial Board to operate measurement stations for the various electrical districts. These Boards have a controlling say on the exploitation of water resources (e.g. control on how EDF dams are managed).

Some organizational improvements being undertaken in the near future are intended to reduce the number of bodies involved.

2.3.3   Use of the results

Before it can be applied in an actual case, the long series of data providing the measurements of flow over several years must be processed by methods appropriate to the analysis required for each particular case. Statistical analysis is particularly important.

Nevertheless, even assuming that the precipitations are independent of each other, the water flows on a given day are not independent of those on preceding days because of the inertia of the catchment area. This correlation decreases as the time interval between the measurements increases.

The inertia of the catchment area depends to a large extent upon the ability of its soil and subsoil to store water and then to return it to the river. The inertia can vary from several days for a small impermeable area to several months for an extensive basin with large storage volumes in the form of underground aquifers, glaciers, etc.

This storage can play a regulating role on the flow of a watercourse. The degree of control is a maximum for underground aquifers and a minimum for storage as snow and ice because of the sudden restitution of stored water with the thaw.

(a)   *Some definitions*

(*Note*: Arithmetic means are used throughout.)

Natural flow is the original flow before construction work commenced or the flow which is reconstituted from measurements made after the insertion of the dam, water intake, etc.

Mean daily flow is the average of measurements over one day. In a period of spate or for streams in a glacial region, the flow can vary strongly from one hour to the next. The frequency of point measurements should then be increased to define the real value more accurately. In particular, the recording of transient phenomena, such as the rise of a flood, requires measurements to be taken at hourly intervals (Fig. 10).

Mean monthly flow is the average for the daily flows over the month under consideration.

Annually averaged mean monthly flow is the average of flows read during the same month for each year of observations.

Mean annual flow (sometimes called the *modulus*) is the average of the mean daily flows or the average of the 12 mean monthly flows weighted by the number of days in each month.

The notion of the average year does not reflect the variation in flows between the different years. It is advisable, therefore, to complete the study of the flows for the average year with at least an analysis of those of a dry year and a wet year characterized by a frequency of appearance which, for example, can be that of the driest and the wettest years over a 10-year period.

A better understanding of the pattern of a watercourse is given by reporting on the same graph the flow curves established for different frequencies rather than the single curve of the annual average.

The flow values can be represented by several different types of curve according to the application. In particular, the curve for the cumulative flows allows the minimum capacity for a peak load or storage reservoir to be determined so that a chosen discharge can be maintained over a given period.

Characteristic flow curves can also be used to calculate the capacity of the micro hydroelectric power station that can be installed on a stream (without a regulating dam) and can be used to define the reserved flow. These characteristic curves will be given particular attention in the following sections.

Some particular flow categories have been given a precise conventional definition because of their frequent use. Thus:

(a) The *characteristic maximum flow* (CMF) is the daily flow which is
     exceeded on only 10 days per year (whether consecutive or not). It is

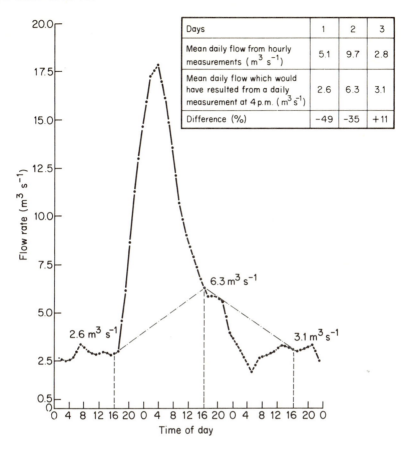

Days	1	2	3
Mean daily flow from hourly measurements ($m^3 s^{-1}$)	5.1	9.7	2.8
Mean daily flow which would have resulted from a daily measurement at 4 p.m. ($m^3 s^{-1}$)	2.6	6.3	3.1
Difference (%)	−49	−35	+11

Fig. 10 Influence of frequency of measurement on the estimation of mean daily flow for the River Sarre at Keskastel (in spate from 3 to 5 August 1970). ———, Observed hydrogram with hourly measurement; — · — · — , hydrogram which would have been observed with a daily measurement taken at 4 p.m.

normally regarded as the maximum flow because the accuracy of measurement decreases for extreme values and the extreme maximum recorded is often not of practical interest.

(b) The *characteristic average flow* (CFA) is the median daily flow, i.e. that which has a frequency of 0.5.

(c) The *characteristic flow for x months* (CFx) is the monthly flow which is exceeded in *x* months of the year whether consecutive or not (frequency *x*/12).

(d) *Characteristic low-flow* (CFL) is the daily flow which is exceeded on 355 days of the year.

(e) *Absolute low-flow* is the lowest daily flow observed (this value is of little practical use).

Summary 5   Practical method for manual determination of the characteristic flow curve for the winter period (1 October–31 March).
Example: Monthly flows in cubic metres per second for the River Ellé at Pont ty Nadan, Brittany (catchment area = 580 km²), for 1958–1974

1. Classification in increasing order of the six monthly flows for each winter period (classification by row) (tabulation 2).
2. Classification in increasing order of the n first to n sixth values; n = number of years = 17 (classification by column) (tabulation 3).
3. Interpretation of the results:
   (a) The first column of tabulation 3 gives the flows which are guaranteed for six months out of six.
   (b) The second column gives the flows which are guaranteed for five months out of six.
   (c) The kth column gives flows guaranteed for $6 - k + 1$ months out of six.

Each column shows n observations (one for each year) of each characteristic flow. Each characteristic flow varies from one year to the next. It forms a *population* of n values ($n = 17$). This population can be characterized, for example, by (a) its median or central value and (b) the value guaranteed for x years out of 100. If n is the number of years, the value which is guaranteed for x years out of 100 is the value of the ith column such that

$$i = (1 - (x/100)) \, n + 0.5.$$

Example: For the flow guaranteed for 5 months out of 6, and 8 years out of 10 (or 80 years out of 100),

$$n = 17; \quad x = 80.$$

The column number is 2 and the row number is

$$i = (1 - (80/100)) \, 17 + 0.5 = 3.9.$$

Thus, the answer is found by using a weighting of 90% of line 4 and 10% of line 3; i.e.

$$Q = (2.5 \text{ m}^3 \text{ s}^{-1} \times 10\% + 3.9 \text{ m}^3 \text{ s}^{-1} \times 90\%) = 3.76 \text{ m}^3 \text{ s}.$$

Similarly, for the whole winter:

	No. of guaranteed months			
	6 months out of 6	5 months out of 6	4 months out of 6	3 months out of 6
Median	3.4	6.3	8.3	13.5
8 years out of 10	2.0	3.8	6.0	9.9

Tabulation 1   Winter

Year	Jan.	Feb.	Mar.	Oct.	Nov.	Dec.
1958	19.3	23.1	23.9	10.0	6.8	17.9
1959	35.8	10.6	8.7	2.3	5.4	35.9
1960	31.7	18.9	24.6	20.5	41.8	24.3
1961	30.3	11.1	8.1	2.2	3.9	23.4
1962	8.8	11.3	27.2	3.4	5.6	6.0
1963	7.4	8.0	24.1	1.9	2.3	19.3
1964	17.9	37.1	15.0	3.8	12.0	7.4
1965	17.7	25.2	19.7	7.9	9.4	38.8
1966	22.1	10.0	4.5	6.4	13.9	15.5
1967	17.4	12.8	15.7	4.5	5.0	11.5
1968	15.1	26.2	13.9	1.1	4.8	10.0
1969	12.6	10.9	9.0	2.0	7.4	8.1
1970	8.3	19.5	13.5	0.8	1.6	6.3
1971	12.1	12.5	7.0	2.5	8.0	2.5
1972	17.2	35.9	14.6	4.6	2.4	16.5
1973					14.9	5.6
1974						13.0

Tabulation 2   Classification in each row (winter)

Year	1	2	3	4	5	6
1958	6.8	10.0	17.9	19.3	23.1	23.9
1959	2.3	5.4	8.7	10.6	35.8	35.9
1960	18.9	20.5	24.3	24.6	31.7	41.8
1961	3.7	3.9	8.1	23.3	23.4	31.7
1962	3.2	5.6	6.0	10.4	11.1	30.3
1963	1.9	8.8	11.3	19.3	22.3	27.2
1964	3.8	8.0	7.3	7.4	12.0	24.1
1965	7.9	9.4	11.1	12.0	13.9	38.8
1966	6.4	11.5	15.0	15.5	34.9	37.1
1967	4.5	4.5	13.9	17.7	19.7	25.2
1968	1.1	4.8	5.0	10.0	10.0	22.1
1969	2.0	6.3	7.4	12.8	15.7	17.4
1970	0.8	1.6	2.5	13.9	15.1	26.2
1971	2.4	8.0	5.6	13.5	16.5	12.6
1972	4.6	13.0	14.6	14.9	12.1	19.5
1973					17.2	35.9
1974						

Tabulation 3   Classification in each column (winter)

Row	1	2	3	4	5	6
1	0.8	1.6	2.5	7.0	7.4	12.5
2	1.1	1.9	5.0	7.4	10.0	12.6
3	1.9	2.5	5.6	9.0	10.9	17.4
4	2.0	3.9	6.0	10.0	11.1	19.5
5	2.2	4.5	7.3	10.4	12.1	22.1
6	2.3	5.4	7.4	10.6	15.1	23.9
7	3.4	5.6	8.1	12.8	15.7	24.1
8	3.7	6.3	8.1	13.5	17.2	24.2
9	3.8	8.0	8.7	13.9	17.9	26.2
10	4.5	8.8	11.3	14.9	19.7	27.2
11	4.6	9.4	13.9	15.5	22.3	30.3
12	6.4	10.0	14.6	17.7	23.1	31.7
13	6.8	11.5	15.0	19.3	23.4	35.9
14	7.9	13.0	17.9	19.3	31.7	37.1
15	18.9	20.5	24.3	24.6	35.8	38.8
16						41.8
17						

(b) *Formulation of characteristic flow curves*

Measurements over a period of at least 10 years are required for the construction of these curves if they are to be of use for micro hydro-power stations. If records are not available for a sufficient number of years, they have to be extrapolated to the minimum 10-year period by using, for example, rain-flow simulation models or by correlation with measurements at neighbouring stations with a sufficiently long series of recorded measurements.

Strictly speaking, the characteristic flow curves should be constructed so as to cope with the problems which can arise during high- and low-flow periods. These include topics such as spillway sizing and definition of the reserved flow.

On account of the large volume of data involved, a computer is required. However, in a preliminary study for catchment areas of at least 100 km^2 in an oceanic climate, the curve can be determined manually using the monthly average flows for the years being studied (see Summary 5).

In general, this simplified method introduces only a very small error except for the extreme values which have to be examined separately. In particular, the characteristic low flow cannot be determined by this method.

(c) *Accuracy of the calculations for characteristic flows*

The accuracy of the results in Summary 6 depends upon three parameters: (a) the particular characteristic flow required (the average flow for a year, 6 months, 3 months, or 1 month); (b) the number of years of observation; and (c) the pattern of the watercourse as characterized by the variability of the flow and expressed by the standard deviation of the characteristic flow.

The accuracy in the determination of the characteristic flow increases as its variability decreases and as the number of years of records increases.

(d) *The problem of short-series data*

A catchment area of tens or hundreds of square kilometres is an element of a more general climatic, morphological and local geological context, which occasionally permits the flow results for one catchment area to be transferred to another neighbouring one. This is very useful where the site studied has not been monitored or where only a few years of flow measurement are available.

Summary 6 Calculation of the accuracy of a characteristic flow

1. Calculation of the mean m of n observations (x_i):

$$m = \frac{1}{n}\, \Sigma x_i.$$

2. Calculation of the sum of the squares of each observation x:

$$S_2 = \Sigma x_i^2$$

3. Calculation of the standard deviation: this function is available on some pocket calculators:

$$s = \sqrt{\frac{S_2 - nm^2}{n - 1}}$$

4. Calculation of the range for a confidence level of 80%: the calculated quantity d is the interval on either side of the calculated value which retains a confidence level of 80%.
   (a) For the median, $d = 1.28s/\sqrt{n - 1}$.
   (b) For a flow guaranteed for 8 out of 10 years, $d = 1.49s/\sqrt{n - 1}$.
   (c) For a flow guaranteed for 9 out of 10 years, $d = 1.73s\sqrt{n - 1}$.
5. Graphical representation: example of the variation at an 80% confidence level of the median flow for the winter period of the River Ellé (the flow which is guaranteed for 3 out of 6 months) according to the number of years of data.

n first years	q median	m	s	d
3	10.6	12.8	8.84	8.00
5	19.3	17.64	6.80	4.35
7	19.3	16.4	6.90	3.60
9	15.5	15.8	6.18	2.80
11	15.5	15.5	5.68	2.30
13	13.9	15.2	5.16	1.91
15	13.5	14.6	5.27	1.80
17	13.5	14.19	5.21	1.07

The increase in accuracy with number of years of data is shown in the diagram below:

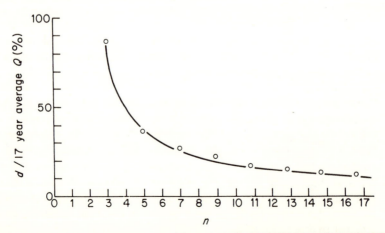

Summary 6 shows the danger of using the flow values directly due to the very large uncertainties involved with the calculation of characteristics based on a short series of data.

Care has to be taken, therefore, to ensure that the catchment areas being compared are similar in size, altitude and relief, hydroclimatic characteristics, vegetation type and geological features.

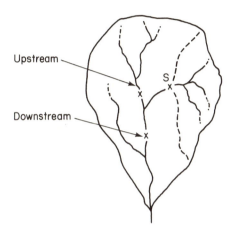

In the last case, if the basin is founded on a karstic limestone, great care has to be taken when comparing basins or extrapolating the flow along the watercourse. In this type of area, sudden increases or decreases in flow can occur because of the substrata. Also, the lack of correlation between surface and underground catchment areas prohibits any comparison between the drained surface areas. In the most favourable cases, comparison of the flows in two or more similar basins can be made, provided that certain precautions are taken:

(i) The flow in a secondary basin, S, should not be calculated by taking the difference between the flows upstream, US, and downstream, DS, as this introduces substantial errors, especially if S is small relative to the upstream basin.

(ii) When a long series of observations is available for one basin, A, a method of comparison exists whereby the monthly flows of basin A are placed on the abscissa and the monthly flows of the neighbouring basin, B, are placed on the ordinate for a common period of measurements. If this approach is adopted, it is necessary to separate the low flow periods from the high-water periods, because, during high-water periods, runoff is preponderant and a good correlation is generally found between different basins as the flow is largely proportional to the basin area.

In contrast, during low flow periods, the flow is chiefly controlled by the intrinsic characteristics of the basins and their supply of ground water. Thus, crude extrapolation will usually lead to serious errors.

(e)   *Extension of the data series – Application to the processing of short series*

Extension of the data series implies the construction of a history of daily gaugings from a known series of daily rainfall records. This requires the representation of the hydrological cycle by means of mathematical equations expressing the balance of the flows. It was shown earlier that the various elements can only be calculated with good accuracy for time periods sufficient to allow the variations in the water storage to be neglected.

The time step used here (1 day) does not allow these variations in storage to be neglected, and the complexity of the phenomena will again lead to the use of empirical relations which cannot be justified by physical considerations and whose parameters will not relate directly to those which characterize the catchment basins.

Nevertheless, even if the physical description of the phenomena cannot be complete, the empirical relations are not entirely devoid of physical meaning. They seek merely to give an overall picture using phenomena which are simpler but similar in type. The ground, for example, will be treated as a succession of reservoirs simulating the water retention mechanisms at different levels as temporary retention by vegetation, the soil and the subsoil.

Each retention level can transfer water to the levels immediately above or below according to simple mechanisms whereby, for example, the water in a deep reservoir is not available for evapotranspiration because it is not in a capillary relationship with the topsoil which is penetrated by the roots of vegetation.

The models for the correlation between precipitation and flow seek to aggregate the various phenomena introduced by the drainage basin, by looking for what happened to the various precipitations on traversing the geological filter.

A direct correlation between precipitation and flow for a single drainage basin cannot be identified, however, as the geological filter formed by the basin introduces a delay, or a distortion of the response, which depends upon previous precipitation, whether it occurred several days or several months earlier. The geological filter, therefore, has a 'memory' which makes direct correlation very approximate and tentative.

The nature of the filter is not accounted for by the analysis, and the models are only involved with the input and output, as the filter is simulated by artificial parameters which cannot be measured directly on the ground. It is necessary, therefore, to have a reference series, which in this case will be a history of the recorded daily flows, with the objective of adjusting the values of the parameters to relate the flows as accurately as possible to the precipitation over the same period. In practice, the optimization should be performed with precipitations which include the period before the first measured flows so as to allow for the memory effect of the basin. This adjustment is carried out during the calibration phase of the model (Fig. 11).

Next, a precipitation history over a long period of time is supplied to the calibrated model which then produces a history of the daily flows for the corresponding period. This flow history is then analysed by the usual statistical methods.

The accuracy of the results from a data-extrapolation model is much better than that of a classical statistical treatment which relates only to a few years of data. It has already been shown that the range to yield a confidence level of 80% for the median flow for the winter period (6 months) for the River Ellé was 11.88/13.5 = 88% for three years of observations and 39% for five years of observations.

The use of the calibrated simulation model for one single year of observed flow, by using 18 years of precipitation, gives a range at an 80% confidence level (according to the calibration year used) equal to 5% of the median flow for the winter period.

In conclusion, when a short data series (1–7 years) is used, it is better to extend the flow data by correlation with the precipitation rather than to extrapolate the results obtained for neighbouring basins.

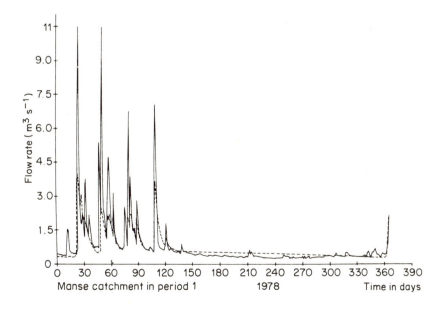

Fig. 11 Reconstruction of the average flow values in a stream from models relating rainfall and flow

The major advantage of using such a short series is that it directly represents the particular basin. Also, the search for correlations between precipitation and flow frees the investigation from any particular dependence on the characteristics of the basin as these are already included in the flow measurements.

This method does not require site measurements to be made as it is based on existing hydroclimatic data. It is therefore less expensive, requiring only a few days' work, and it provides enough information to form a basis for any study of project viability.

If no flow data are available, the period during which the micro hydroelectric station project is being constructed can be used. This will usually allow one year of daily measurements to be collected and processed by the method outlined above.

(f)   *Extremum values: low flow and flood*

The extremes of the characteristic flow curve represent periods of low flow and flood. These terms have acquired particular significance and are defined in terms of a volume of flow and a frequency of occurrence. The objective is to distinguish the ordinary from the extraordinary and to allow values to be chosen which will be more representative of the annual variations.

Thus, the extreme or absolute low flow, which represents the absolute minimum in the known flow, is only of subsidiary interest, and low flow will be better represented by the mean characteristic low flow, defined as the mean flow which is exceeded on 355 days per year in each year over all of the years of observation. For flood conditions, the mean characteristic flood is similarly defined as that flow which is exceeded on only 10 days per year in each year over all the years of observation.

The consequences of extreme flows differ according to their degree of continuity and the proposed use of the electricity produced. Although low flow and flood are very different phenomena, the drop in electricity production that they entail is of more importance to isolated networks than to generators supplying the national grid, which has other sources of production. Also, there is the question as to whether the number of days of low flow and flood should be considered as a loss of generation depending only upon the number of days, or whether it should be related to the continuity of the days considered. That is, are disturbances involving 10 consecutive days of poor generation equivalent to 10 separate days distributed over a longer period of time? As low flow and flood are fundamentally different, it is advisable that they should be examined separately.

Low flows represent the supply of a watercourse from the ground water. They are related to the term B in the balance equation and allow the administrating services to define the minimum required flow, the reserved flow. They also allow an independent producer to evaluate the minimum potential of his site, to determine the periods during the year when it occurs, and to establish whether it is compatible with the proposed use, such as central heating in winter, the operation of an isolated network, etc.

Flood flows are produced by the sudden introduction of runoff (the term R in the balance equation) and a knowledge of their characteristics is needed for sizing the civil engineering works. These will include flood protection for the installation allowing for the maximum rise in the water-level and sizing of the spillway so as to avoid flooding the installation or the area upstream of the intake.

In the case of low head systems ($H \leqslant 5$ m), a flood can also eliminate or greatly reduce the head by raising the downstream water-level. In this case, in spite of the presence of a large flow, the reduction in head greatly reduces the power produced, even to the extent of eliminating it completely. The number of days of flood which result in low or zero generation is, therefore, a hydrological parameter which should be known. Thus, the processing of flow data alone is not sufficient. It is also advisable to survey the site and to determine the values of the head under different flood flow conditions, either observed or calculated.

Measurement of water depth is relatively easy, either directly by gauging throughout the year, or indirectly by examining the debris deposited on banks or low tree branches.

Extreme flows, therefore, can have two types of effect, general and specific. General effects influence the output of the micro hydro-power station in proportion to the flow. In this context, therefore, extreme flows have no particular significance. Specific effects, however, can have serious particular consequences, either on fish life or on the integrity of the dams which were designed and constructed for characterisic flow values. In these cases, the extreme flows are defined in terms of characteristic low flows or flood flows which represent the objectives required.

Extreme flows must also be regarded in a special way because they are very difficult to assess. The low flows that are measured can be affected by human intervention upstream, e.g. extraction of water for irrigation or for town supply, the contribution of industrial and municipal discharges, or the operation of dam gates. These interferences occur throughout the year but their effect can be particularly important during the low-flow period when the natural flows can equal the phenomena that can alter them.

It should also be noted that 17% of all gauging stations in France have low-flow periods which are affected by external influences to some extent. This percentage can be as high as 28–29% for the Loire, Garonne, and Adour basins.

Floods are equally difficult to measure because they are often not well represented by the usual methods of measurement, though stations on large basins have been specially equipped for the measurement of floods.

2.4   DETERMINATION OF THE USEFUL FLOW: AN EXAMPLE

2.4.1   Effect of choice of minimum operational flow on the production of electricity

As turbines can only operate efficiently over a certain range of flows and as financial constraints do not generally allow multiple turbines to be installed, it is advisable to optimize the exploited range of flows in terms of the required use. The *minimum operational flow* is the minimum flow value which will permit the start-up and operation of the turbine. For most basins in the crystalline region of France (e.g. the Massif Armoricain and Massif Central) or for snow-fed basins (such as in the Pyrenees and the Alps) the periods of low flow, although occurring at different times of the year, are very pronounced when the basin is smaller than 100 km^2.

The value of the minimum operational flow therefore determines the method of use and the number of kilowatt-hours which will be supplied during the year. If the turbine cannot operate during periods of low flow, then the power station cannot be used to provide space heating. Thus, in snow-fed basins there is a large risk of 'outage' during the period from October or November onwards. In contrast, a higher minimum operational flow will allow large flows to be used outside the low-flow periods with an increased power production and an annual energy output which can be greater than would be permitted with a lower minimum operational flow and more regular production. This increase in production reaches a maximum in the case of an independent producer in a non-snow fed basin who sells surplus power to the electricity utility at about twice his generating cost. This will occur during winter rather than summer because that is when periods of high water most frequently occur.

Thus, seeking the maximum production from an installation is often incompatible with independent consumption where a seasonal activity such as skiing is involved which requires regular production with a minimum number of days of non-generation. Ski-lifts require power in the winter but unfortunately that is when the good snowfall areas have very low flows.

Statistical evaluation of hydrological series allows the minimum operational flow to be chosen to suit the intended use, by allowing the total number of non-generating days and the duration of outage to be determined as a function of the choice of turbined flows. This is illustrated in the following example.

2.4.2   Example of the calculation of minimum operational flow

This example will discuss the potential of the Semouse river at Aillevillers (Haute-Saône), with a basin of 97 km^2. The hydrological study shows that the number of days of guaranteed operation with various flows are as given in Table 5.

Table 5 Period of operation guaranteed at different flow rates on the River Semouse
at Aillevillers

(a) Summer period: 1 April–30 September

Operation guaranteed for	Number of days of operation guaranteed during the 183 days of the summer period				
	500 l s^{-1}	750 l s^{-1}	1000 l s^{-1}	1250 l s^{-1}	1500 l s^{-1}
9 years out of 10	115	73	<60	<60	<60
8 years out of 10	156	93	66	<60	<60
5 years out of 10	>173	123	90	75	<60
2 years out of 10	>173	160	126	109	90
1 year out of 10	>173	>173	143	116	100

(b) Winter period: 1 October–31 March

Operation guaranteed for	Number of days of operation guaranteed during the 182 days of the winter period						
	500 l s^{-1}	1000 l s^{-1}	1500 l s^{-1}	2000 l s^{-1}	2500 l s^{-1}	3000 l s^{-1}	3500 l s^{-1}
9 years out of 10	156	105	60	<60	<60	<60	<60
8 years out of 10	162	140	80	67	<60	<60	<60
5 years out of 10	>172	157	137	117	102	75	60
2 years out of 10	>172	>172	162	142	120	105	90
1 year out of 10	>172	>172	>172	147	130	110	95

Source: J. Martin, BRGM (1979).

The number of days of operation is inversely proportional to the nominal flow of the turbine. For example, during the summer period in an average year, the number of days of guaranteed operation is 123 at 750 l s^{-1} but is less than 60 at 1500 l s^{-1}.

With a head of about 2 m, it can be assumed that Kaplan turbines are sufficiently efficient to operate with flows ranging between the rated flow of the turbine and one-half of this value.

Thus, for a maximum turbine flow of 3 m^3 s^{-1}, the possible turbine choices can be shown schematically as

1 turbine operating between 3 and 1.5 m^3 s^{-1} (1),
1 turbine operating between 2 and 1 m^3 s^{-1} (2),
1 turbine operating between 1 and 0.5 m^3 s^{-1} (3),

In practice, on economic grounds, the available maximum effective low power $((P_e)_{max} = 40$ kw) does not justify the installation of two turbines.

The choice of turbine (3) $((P_e)_{max} = 14$ kW, $(P_e)_{min} = 7$ kW) ensures almost continuous operation throughout the year with a very small outage. It is therefore suitable for an isolated system for the production of electricity to provide domestic central heating during the cold season.

Turbine (2) $((P_e)_{max} = 28$ kW, $(P_e)_{min} = 14$ kW) will operate satisfactorily in winter but not in summer.

Turbine (1) $((P_e)_{max} = 42$ kW, $(P_e)_{min} = 21$ kW) will not usually operate in summer but due to its winter production it can supply almost as much energy over the year (about 120 000 kWh in the average year) as turbine (2) (41 000 kWh in summer and 84 000 kWh in winter).

Thus, for the theoretical case of selling the electricity produced to the national grid, turbine (3) with its winter production will offer a larger income than turbine (2) where almost a third of its production is during the summer period, bearing in mind that the winter price is almost twice that in the summer (see Chapter 5).

2.4.3   The reserved flow: its effect upon exploitable flow

The flow discussed so far has been the natural flow which specifies the potential of the site. The installation of a micro hydro-power station on a watercourse has an environmental impact which may sometimes be negligible, but which can be most important when applying for planning permission. This is because the flows during the low-flow period are low in small basins and the interests of the operator may be in conflict with those of other users or with the requirements for fish life in the river.

The *reserved flow*, established by the administrating department when authorizing the installation of the micro hydro-power station, is intended to provide for the various, often conflicting, uses of the water in the river. In practice, this flow is often determined on a 'give and take' basis, taking account of the purpose of the desired installation, of the fauna (whether the fish are migratory or not), and of the rights of the other users.

It is also possible for the low flow to be zero in arid regions, and even in France some rivers with large basins can dry up completely. The Viaur (a tributary of the Aveyron) had a zero flow of 7 days duration at Thuriès (basin area 1050 km^2, modulus 15.1 m^3 s^{-1}) from 31 July to 6 August 1954. The Sarthe has completely dried up, at Spay, on three occasions over the past twelve centuries – around 800 AD in the time of Charlemagne, around 825 AD in the time of Louis le Débonnaire, and in June 1168. The basin area in this case is 5458 km^2.

not fed by ground water and yet they have winter flows of several cubic metres per second.

The choice of the hydrological criteria used to determine the reserved flow is thus of prime importance. Using the value of exceptionally low-flow periods would not be sensible, and practice tends to generalize the use of the *characteristic low flow* (CLF), which represents the average daily flow exceeded on 355 days per year for a certain number of years, with a minimum of at least 10 years.

Thus, very precise rules can be drawn up to represent the particular characteristics of certain regions. For example, in the Auvergne–Limousin region the reserved flow is defined as being between 1.3 CLF_{355} during the warm season (low flow) and 0.7 CLF_{355} during the winter season (1 October–31 March). In this case, CLF_{355} is the 10-year mean of the flow which is reached or exceeded on 355 days per year, whether consecutive or not.

For a particular site, the reserved flow is selected within these extreme values as a function of the length of river between the points of extraction and restoration of the water and of its value for fish life, provided that no other users of the water are involved.

The size of the reserved flow influences the profitability of the micro hydro-power station directly. In the majority of cases, the reserved flow bypasses the turbine and cannot be utilized. The effect over the year is thus concentrated into the period of low flow because, during periods of high water, the available flow normally has greater generating potential. As the interest of the independent producer is to turbine as much water as possible, it would be advisable to make a better study of the characteristics of regional low-flow periods so as to be able to establish more accurate rules for defining reserved flows. These rules are currently too dependent on the balance of power between the various users of the river, and on data which is much too sparse, especially for small rivers where the problem of the reserved flow is most important.

The study of the variability in flows at the beginning of this chapter has shown the length of time needed to obtain a characteristic flow for the river. A few measurements of low flow are insufficient for deciding the values to be given to the reserved flow. On the other hand, determining the reserved flows by using specific flows for stations of long standing but which are sited well downstream of the high basins unjustly penalizes the individual producer by sometimes imposing a flow greater than the actual low flow in the high part of the basin. This, since it is fed mainly by runoff, has normally much more pronounced low-flow periods than the lower parts of the system.

Since a large-scale study of small basins would be required to provide the necessary data and since this could not really be envisaged solely for the planning of micro hydro-power stations, on-site gaugings for one year in combination with the models relating rainfall to flow would seem at the moment to be the most accurate and least expensive method of simultaneously providing the data needed for selecting the options for the reserved flow and of acting as the basis for a serious study of the viability of the micro hydro-power station.

Once defined and clearly stated in the operating specifications, the reserved flow should be respected. It thus implies that appropriate actions should be taken for the management of water use.

BIBLIOGRAPHY AND FURTHER READING

ANONYMOUS (1978). *Distribution de fréquence adaptée aux étiages: méthodes manuelles et graphiques*, Centre technicale Génie rural, Eaux et Forêts, Antony.

BEAUREGARD, J. de (1978). Le bas débits des cours d'eau en France, étiages normaux et exceptionnels, fréquence, répartition. *Bulletin de BRGM, Section III*, no. 3, 215–223.

BODELLE, J. and MARGAT, J. (1980). *L'eau souterraine en France*, Masson, Paris.

BONNET, M., DELAROZIÈRE-BOUILLIN, O., JUSSERAND, C. and ROUX, P. (1970). Calcul automatique de 'bilans d'eau' mensuels et annuels par les méthodes de Thornthwaite et de Turc. *Internal report no. 70 SGN 107 HYD*, BRGM, Orléans.

CARLIER, M. (1972). *Hydraulique générale et appliquée*, Eyrolles, Paris.

CAUVIN, A. and GUERRÉE, H. (1968). *Éléments d'hydraulique*, 7th edn, Eyrolles, Paris.

COUDRAIN, A. and THIERY, D. (1978). Estimation de pluies efficaces avec l'algorithme de Turc. Influence du pas de temps de calcul et de l'utilisation de données moyennes. *Internal report no. 78 SGN 640 HYD*, BRGM, Orléans.

CUINAT, R. and DEMARS, J. -J. (1980). *Débit réservé et autres précautions piscicoles actuellement imposées lors de l'installation de microcentrales électriques en Auvergne–Limousin*, Clermont-Ferrand, Conseil supérieur de la Pêche.

DELAROZIÈRE-BOUILLIN, O., (1971). Utilisation comparée des formules de Thornthwaite, Turc mensuelle, Turc annuelle et Penman, pour le calcul de l'évapotranspiration potentielle et de l'évapotranspiration réelle moyenne. Application au territoire français. *Internal report no. 71 SGN 173 HYD*, BRGM, Orléans.

DUBREUIL, P. – Le rôle des paramètres caractéristiques du milieu physique dans la synthèse et l'extrapolation des données hydrologiques recueillies sur bassins représentatifs. In *Publication no. 96*, International Association for Scientific Hydrology, Washington, DC.

MAZENC, B. (1980). Application de modèles hydrologiques conceptuels à cinq bassins versants bretons. *Internal report no. 80 SGN 015 HYD*, BRGM, Orleans.

REMENIERAS, G. (1972). *L'hydrologie de l'ingénieur*, 3rd edn, Eyrolles, Paris.

THIERY, D. (1977). Calculs de pluie efficace au pas journalier avec sous-programmes CLIMAT et CLIDAT. *Internal report no. 77 SGN 211 HYD*, BRGM, Orléans.

THIERY, D. (1979). La prévision appliquée à l'hydrogéologie. Unpublished document, BRGM, Orléans.

THIERY, D. (1980). Détermination du debit d'équipement d'une microcentrale hydro-électrique. *Internal report no. 80 SGN 405 EAU*, BRGM, Orléans.

CHAPTER 3

The planning and construction of micro hydroelectric power stations

In general, micro hydroelectric power stations (MHPSs) take their water from low dams. Suitable sites for MHPSs occur naturally where there are favourable land features such as travertines, natural rock barriers formed by glacial or volcanic flows, and points where a valley narrows. Mountain lakes can also provide suitable sites if their outlets can be dammed or if they can be tapped at a lower level after tunelling. Alternatively, dams of former mills and river-locks are ideally suited to conversion to micro hydroelectric applications.

3.1 SITE WORKS NEEDED FOR A HYDROELECTRIC INSTALLATION

A hydroelectric installation normally comprises the various elements which were indicated in Fig. 1. These include the following:

(a) The works on the river (usually a low dam) to divert the water towards the site where energy is to be produced while maintaining the minimum reserved flow. Sometimes a reservoir is also created. Depending upon the morphological and topographical conditions, the intake can be (i) a rigid concrete dam, a flexible dam of earth or rockfill, or a dam composed of adjustable shutters mounted on piles and concrete aprons; (ii) a more rudimentary dam constructed from gabions, masonry, a system of planks and blocks, an underwater intake, etc.; (iii) an existing diversion constructed to direct water towards a mill or irrigation channel.

(b) A spillway to protect the various siteworks by draining floodwaters whose frequency and flows were defined in the hydrological study (Chapter 2). Particular attention should be paid to the sizing and construction of the spillway for earthfill dams so that they can withstand the force of the discharge, although their cost must remain compatible with the aims of the project and with the income from the energy production.

(c) Arrangements for the passage of migrating fish (Chapter 7) to protect the aquatic fauna by countering the effects of the permanent dams. These structures, generally called fish-passes, must be designed and built on the basis of a good knowledge of the directions and times of migration of the various species populating the rivers.

Under the French Law for the Protection of Nature (1976), the contracting authority is responsible for installing suitable measures to alleviate the effects of all types of environmental impact.

(d) *The head-race*: The full-pipe bringing the water to the turbine can be placed directly on the intake, but, depending upon the inlet conditions or the application, it may be necessary first to provide an earth or concrete channel. This feeder has a low gradient and comprises the following:

(i) At its head, a grating or grill with cleaning scraper to trap material such as leaves, branches, or pebbles which are washed along by the river. A system for closing off the water intake should be installed for maintenance of the various works.

() (ii) At its end, where the full-pipe begins, a settling tank and a discharge chamber, or forebay. It is essential that the forebay is provided with an overflow dam and spillway to ensure a bypass flow when the power station is not operating.

The discharge chamber is introduced to compensate for any small differences between the flow diverted to the powerhouse from the river and that required for the turbines.

(e) *Full-pipes*: These circular pipes can be of different types, depending upon financial conditions and the facilities of local industry. They are usually made of steel, rolled sheet, PVC or polythene. To protect the full-pipe, it is often necessary to add a balance chimney or to site a surge tank as close to the turbine as possible.

Installation in areas subject to earth movements or avalanches should be avoided. Steep slopes should also be avoided because of the heavier construction and maintenance costs that they involve.

Finally, particular care and attention should be paid to the visual integration of the works with the surroundings. Some topics requiring particular care are planting, the choice of colours, ground disturbance, etc.

(f) *Gates and valves*: Equipment for closing the water pipes must be installed. The type of device used depends on the pressure and the allowable losses. Gate valves, butterfly valves which can be used up to a pressure of 20 bars or ball valves which introduce no load losses when open can all be used.

(g) *Buildings*: Lightweight buildings are constructed to protect the electromechanical equipment, and, in addition to the usual norms, they should allow for the possibility of flooding and for their visual integration with their surroundings. Particular care should be taken with the power lines, which may be buried. A silencer can also be installed, depending on the proximity of dwellings.

(h) *Tail-race*: At the turbine outlet, the water is returned to the river by a tail-race which is dressed at its upper part to avoid undermining or damage to the power station. It is necessary to include a grill to prevent entry by fish migrating upstream.

3.2 Criteria for site selection

The choice of site is based on a close interaction between the various conditions – the pattern of the river, the integrity of the site works, environmental integration, and the conditions defining the costs and benefits.

Having established an inventory of energy demands in an area with known hydro-potential, the various parameters must be evaluated. It is important, however, to bear the other possibilities in mind as the hydraulic resources should be assessed simultaneously with those from various other sources, such as geothermal (high and low enthalpy), heat pumps (water-to-water, water-to-air, etc.), biomass, solar (from panel or photovoltaic cell) and wind.

The factors involved can be classified in terms of

(a) *The head*: Fundamental data for evaluating the power and the output. The loss of head and rate of fall of the river through this head loss allow the output to be calculated for a given length of pipe.

(b) *Hydrological pattern*: Defined from measurements or from the relationships between effective rain and discharge (Chapter 2). This is essential for calculating the size of the works required for the intake and for the production of energy, and for forecasting the profitability of the operation.

(c) Usage of the water, upstream of the intake to determine the flow which is available, and downstream to determine the effects of diverting the water from present and future uses.

(d) Environmental impact of different installations and the countermeasures to be adopted: noise, aquatic flora and fauna, aesthetic factors (see Chapter 7).

(e) Size of the works involved and evaluation of their stability depending on the various lithological, morphological, topographical and geotechnical conditions.

(f) Distance from the intake to the power station and from the power station to the consumer site (to calculate the length of the electricity lines).

The decision to proceed with the installation is reached on the basis of a cost–benefit analysis carried out at various levels during the planning stages, with the degree of refinement of the analysis depending on the type of work involved and the energy output.

3.3 Preliminary geological considerations

Various geological events can affect the sides and floors of valleys and can damage man-made structures. Many examples of the type of geological event which can affect a hydroelectric installation have been reported in specialist publications. Some typical examples are

- (a) A landslide which obstructed a valley (Isère, Pontamafrey)
- (b) Damage to the turbine through rapid silting at an intake (Nepal)
- (c) Leaks in fissured sandstone which prevented dams from providing the desired water level (Saint-Guilhem-le-Désert (Hérault), and small installations in Jamaica)
- (d) Expensive reinforcement works being required on a new construction site when it was discovered that alluvial deposits were hiding an earlier landslide (the dam at Tark on the Sefi-Roud in Iran)
- (e) Earth dams damaged due to incorrect sizing of spillways (at Wadi Ghan, Libya)
- (f) Destruction of installations with foundations in gypsum soils (as occurred in a canal on the Vésubie in the Alpes-Maritimes, and in some irrigation canals of the Euphrates in Syria).

In a survey carried out in 1973, the International Committee on Large Dams reported that, over the period 1900–65, 466 dams throughout the world had been affected by some type of incident. Incidents relating principally to stability, the effects of surface water, either underground or inside the works, or to flood drainage, have affected 37 dams with heights of 5–15 m; 218 with heights of 15–30 m, 110 with heights of 30–50 m, and 37 with heights of 50–100 m.

The diagnosis of the geological formations involved, which are usually caused by surface weathering, should be followed by a low-cost reconnaisance study. The geology applied, although apparently rudimentary, cannot be properly understood without a firm basis of fundamental geological principles. The opinion of the geologist, the recommendations that he draws up, and their subsequent institution form the guidance which the contracting authority or the master of works should follow in order to avoid the risk of destroying the works, or at least of having to install expensive extra works which would eliminate any financial benefit of the project.

3.3.1 The geological approach

Various elements of the geology of the area must be considered. These include the structure, type of rock, weathering, erosion, transport and the danger of earth movements for the whole catchment or for the sites of individual elements of the micro hydroelectric power station itself, and in particular for the intake works.

The origin of the rocks, which explains their mineralogical nature, and the detection of tectonic movements which have influenced their deposition, result in the definition of their weak points, their susceptibility to weathering, their evolution over time and, finally, their reaction to artificial interference. The water causes weathering by

(a) Acid attack of the minerals in the rock (e.g. with water containing carbon dioxide gas). This phenomenon is gradual in a temperate climate but is quite severe in the tropics. It leads to the formation of clays.
(b) Solution of carbonate and saline rocks.
(c) Hydration of the anhydrite, which is the basis of the formation of gypsum with accompanying swelling; hydration of lightly hydrated ferric salts such as haematite or goethite produces hydrated ferric salts such as limonite with an increase in volume as seen, for example, in weathering of the Vosges sandstone.
(d) The oxidation of iron (ferrous salts), which results in its liberation and consequent loss.

The phenomenon of weathering is amplified because water penetrates the fissures created by discontinuities in the rock mass through stratification joints, diaclases, faults, etc. Variations in temperature will accelerate the degradation of the rock.

Rock which is broken down in this way will be particularly sensitive to the effects of erosion, a phenomenon which only begins when small streams of water have combined and acquired speed and hence sufficient destructive energy.

Running water is an essential factor in the shaping of the relief of a continent. The scale of the erosion depends on the degree of weathering and fragmentation of the rocks. The remaining formations will be carried away and redeposited elsewhere to a degree which depends on the energy built up by the currents.

A good example of this is the evolution of limestone which can be eroded into sculptured shapes and gorges at the surface, and grottoes, limestone caves, and galleries inside the mass (e.g. karstic morphology).

Glaciers are also agents of erosion and accumulation. They were common in the early Quaternary, but their action is now very much reduced or at least restricted to very high valleys.

Weathering and erosion together with the transport and accumulation of sediments form a series of phenomena which have united to create the present mineral landscape, on which vegetation and sometimes man-made constructions have been superimposed.

3.3.2 Identification of sensitive zones

Geological formations react in different ways to the action of water depending upon their lithological nature and their degree of weathering. Their stability and movement, over both the long and short terms, should be considered when investigating sites for micro hydroelectric power stations. It is important to evaluate all the geological risks which could compromise the structure.

At the same time, during the geological evaluation, the areas which are subject to avalanches should be identified even if they have not been affected for decades.

It is the geologist's task to decipher the various hidden factors involved by interpreting aerial photographs for the basin and then by a more detailed examination on site.

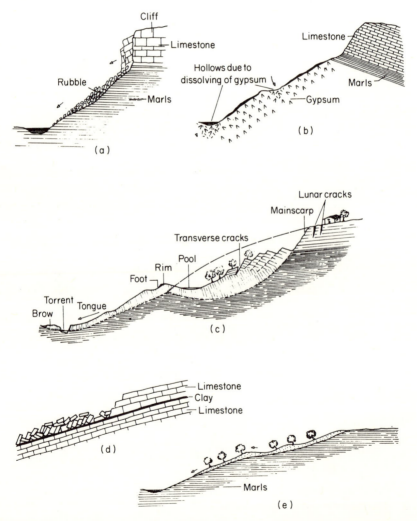

Fig. 12 Earth movements which could affect micro hydroelectric installations: (a) collapse; (b) subsidence; (c) landslide (rotational); (d) landslide (plain); (e) creep-solifluxion

(a) *Earth movements*

Earth movements can occur with little notice and can affect wide areas. They can also develop slowly, but, in all cases, structures sited on their passage will be damaged to some extent.

Among the main types of earth movement (Fig. 12) are falls of rock and boulders, collapse of rocky masses, subsidence and caving in, landslides and mud-flows.

In France, the problems created by earth movements have been studied since 1972 under a programme (ZERMOS) dealing with zones exposed to surface or underground movement. The national geological department is responsible for drawing up Zermos maps, which identify those unstable areas which are subject to earth movement.

The Zermos map assesses the probability that a certain type of earth movement will occur for a given region, i.e. that a particular instability can be completely excluded over a certain time period. The stability of sites and their degree of vulnerability to displacement can be determined from the information on these maps. They include such natural factors as the lithology, structure, drainage, gradients, history of soil movement, and vegetation, and they also indicate human interference through building activities and road networks.

It is worth noting that up to now these 'warning' documents have not had the same regulatory value in France as the corresponding recommendations have had for areas subject to avalanches, but, where they exist, they do form a constraint which should not be ignored by local authorities or other developers. They are also taken into consideration in outline planning and town planning decisions under the various development plans.

A law promulgated in July 1982, providing for the payment of compensation to victims of natural disasters, has given rise to new types of map for sensitive areas of French territory. These maps are intended for use by insurers, public safety officers, and mayors who will be responsible in the future for the issuing of building permits.

Countermeasures   Whenever the safety of people or property is at stake, it is important to ensure the stability of the site. The approach adopted should be suited both to the site concerned and to the financial context of the scheme.

Large-scale works should not be undertaken for a micro hydroelectric power station as they are only relevant as part of a regional scheme. The following measures for the stabilization and consolidation of a site are appropriate and can therefore be carried out as required in each case:

(a) *Modifying gradients*: by lightening the top of the slope, smoothing its shape, terracing or spreading earth at the foot to form a thrust or counterweight (provided that this does not contribute to further earth movement). Boulders can be piled as a counterweight at the foot of the slope, and this provides more drainage at the lowest slippage point (again, providing that it does not contribute to further earth movement).

(b) *Drainage*: Ditches can be dug in stable areas to prevent earth movement being initiated by accumulated water.

(c) Planting a limited amount of vegetation, starting with low plant cover such as grass and then proceeding to reforestation. It is sometimes possible to reforest beyond the existing forested areas.

(d) Filling in fissures and cracks with clay materials (grouting).

(e) *Deep drainage*: Ditches dug to a maximum depth of 3–5 m to cope with high groundwater levels (such as high water-tables), as determined by sampling tubes equipped with pumps. Deep horizontal drains are required when landslips are extensive.

(f) *Supports*: Drained supporting walls, using the reinforced earth technique in which reinforcement is combined with the building materials and provides drainage for potentially harmful water. Gabions or rockfill can tolerate a certain amount of deformation.

Some techniques are too expensive in terms of the economics of an MHPS, and if procedures such as injection, bolting, or other larger-scale measures for dealing with any frequent severe torrents are required, the project should be rejected.

(b)   *Avalanches*

Avalanches (Fig. 13) are more frequent than earth movements, which they sometimes cause. They often occur regularly in certain high risk areas, but

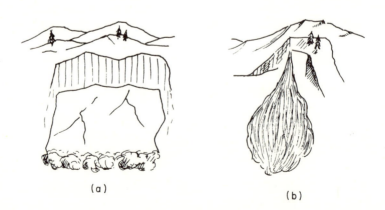

(a)

(b)

Fig. 13   Avalanches: (a) sheet – compacted snow; (b) pear – fast, dry or powdery non-compacted snow, or slow, wet snow

some corridors have not been affected for several decades so that their catastrophic effects have been forgotten. They can be classified as follows:

(a) *Dry snow avalanche*: powdery snow, sliding powdery snow, blown drifts (packed windblown snow)
(b) *Wet snow avalanche*: an avalanche in the form of a sheet of wet snow, balls of snow (occurring in springtime) or melting snow (rotten snow).

Avalanches are often complex in nature. A wind-blow sheet of snow, for example, can set off an avalanche of powdery snow. The uprooting of pebbles and boulders entrained with the snow adds to the damage.

Maps of France similar to the Zermos maps, showing the probable locations of avalanches, have been drawn up. They coordinate all known avalanche data and are drawn on a scale of 1 : 20 000. They are designed for use by local authorities, and can be consulted by private individuals. These documents have no legal power, although they do have a restraining influence on the local authorities. So far, the programme has only been completed for the avalanche-prone areas of French mountains.

For safety reasons, an application for planning permission in avalanche-prone areas must be accompanied by maps drawn to scales of 1 : 2000 and 1 : 5000 if urban development is planned or envisaged in the area. The latest revision of the procedure has retained these maps of avalanche-prone areas, but in a new form adapted to the requirements of the various users.

Countermeasures: If construction in vulnerable areas is unavoidable, permanent protection measures must be installed to prevent or deflect an avalanche. Such preventive measures consist of

(a) Deflecting the wind and avoiding accumulation of snow in unstable areas by installing windshields or roof-gutters
(b) Retaining the snow, using banks, terracing, racks, nets, anti-avalanche walls, or gabion walls, always bearing in mind that, in the last two cases, bad planning would lead to the displacement of more material if an avalanche did occur
(c) Stabilizing the snow with vegetation suited to local conditions.

Constructions such as tunnels and galleries to protect or deflect the snow are only introduced when the road network is involved.

The cost of preventive measures is very high, and in 1970 the installation of racks on gradients of between 60 and 120% was estimated at between 300 000 and 500 000 FF per hectare. This type of work is only possible when fully justified by the financial return on the installation.

The preliminary geological examination, which is conducted differently depending upon whether the whole basin or only the site is considered, results in a diagnosis which identifies the steps to be taken for producing a good installation well suited to the needs of the site.

If this approach is carried out at the appropriate time and combined with the energy production possibilities, as expressed by the head and discharge rates, it allows a site to be selected or rejected. It is a first stage and is followed by detailed studies of the various elements of the MHPS installation.

3.4   SITE, FOUNDATIONS, ANCHORAGE AND STABILITY

As MHPS installations do not involve a large financial outlay, the reconnaissance studies will be oriented towards removing any doubts which remain after the surface geological study. The objective is to locate under the best stability conditions the intake, the head-race, the full-pipe and the building housing the electromechanical equipment. It is especially necessary to avoid the construction of a dam which is not suitable for the local geological conditions or the siting of a channel in an area subject to landslides.

The intake on even a shallow watercourse, with a depth, for example, of 2–8 m, involves a dam intended to raise the water-level. As such, it represents a foreign body to be rejected by the natural environment. Also, if it is to retain water, then it must not slip on its foundations; it must not be overturned; it must support exceptional discharges without being damaged; it must not be undermined; etc.

3.4.1   Coping with geological conditions – mechanical properties

At the site, examination of the geological conditions should be carried out during a low-flow period, when more of the surfaces will be revealed.

It is essential to determine the depth of the foundation soils down to the substratum (as indicated by its hydraulic and mechanical properties). The geological character of the soils and their hydraulic and mechanical properties must be defined, and those cases where the topsoil is to act as the intake bed must be identified.

Studies should then be undertaken on different aspects depending on the preliminary geological survey. In some cases, investigation by such rudimentary means as a pick, shovel, auger, etc., will be sufficient, but it will sometimes be necessary to use more advanced techniques involving geophysics, drilling, water and other samples, laboratory tests, etc.

The data collected by experiments on the site and in the laboratory will first be used to calculate the foundations required, i.e. to predict the forces of constraint and the allowable deformations which would avoid any irreparable fracture of the structure, then to specify the anticipated degree of permeability of the bed to estimate the tolerable leakage rate when considering the constraints on the stability of the site works and on requirements for energy production.

Generally, the reconnaissance work involves a geophysical survey which is determined from lithological data collected by mechanical samplings, water tests for permeability, mechanical tests on the site, laboratory investigation of the mechanical properties of the soils and possibly of the rocks.

As a general rule, investigations on the site and in the laboratory can only be carried out to the degree of detail appropriate to the available geological data for the area and within the desired financial limits.

3.4.2 Water intakes

Given a greater or lesser knowledge of the geological and mechanical properties of the site and of the interstitial and underground movement of the water both for the site and the surrounding area, the construction of the MHPS can begin with the intake. Its size will depend on the available head.

If the head is high, a simple intake allows the water to flood into the full-pipe and is protected against solid material in the flow by installing grills or a settlement tank, as with the submerged intake.

If the head is low, the intake will comprise a dam to raise the level of the water flowing into the feed pipe. This is more difficult to achieve as, in most cases, the cost of the project is to be low. It is important either to use natural barriers, such as travertine deposits or volcanic flows, or man-made ones such as mills or lock-gates, which can be employed to give better economics.

The MHPSs intakes can have gravity, earth or rockfill dams. More rarely, arched vaulted dams are employed, and, depending upon local conditions, other types can also be used.

The stability conditions, i.e. the conditions needed to avoid the dam being turned over or sliding on its foundations, can be analysed using a number of simplifications.

(a) *Gravity dams*

Concrete dams are usually triangular in section and their stability depends upon the forces present (see Fig. 14).

Weight of installation, $W = (hb/2)\gamma$

Water pressure, $P = (h^2/2)\gamma_w$

Under-pressure of infiltrated water, $U = (hb/2)\gamma_w$

Resultant of the forces, R

Density of water, $\gamma_w = 1 \text{ t m}^{-3}$

Density of concrete, $\gamma = 2.4 \text{ t m}^{-3}$

Angle of friction, ϕ

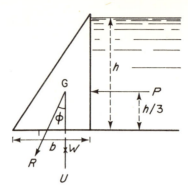

Fig. 14 Stability of a gravity dam

Stability against being turned over usually implies that the resultant of the forces passes through the central section of the base of the dam.

Resistance to sliding, according to French safety regulations, implies the condition $P/(W - U) < 0.75$ (where 0.75 is the concrete–rock coefficient of friction corresponding to the angle $\phi = 37°$. To counteract sliding, U should be decreased by reducing the permeability upstream of the wall (or by using a combination of concrete and rocks) possibly accompanied by drainage.

A concrete dam can be sloping. If the depth of the water is greater than the height of the dam, stability will be ensured by using a larger structure and making it trapezoidal in shape. A concrete dam can be fitted with gates which open to allow floodwater to pass through (see Fig. 15).

When constructed of concrete, various types of joint should be installed:

(a) Expansion joints, which are intended to avoid cracking the concrete when the intake is closed or with variation in temperature
(b) Water-tight joints installed with the expansion joints. These flexible, water-tight joints are made of sheets of metal, such as copper, lead or zinc, or of plastic, such as polyvinyl chloride.

(b) *Dams of earth or rockfill* (Fig. 16)

These structures are formed of loose material available on the site. They are more flexible than gravity dams and weigh four to ten times more than a concrete dam. Flows to the inside of the structure are allowable provided that they are monitored and controlled. Particular care should be taken in defining the stability conditions for the embankments.

The flow through a loose, homogeneous material which is sufficiently water-tight is distributed over a set of parabolic flow surfaces. Seepage can occur on the downstream slope for a quarter to a third of its height, with the

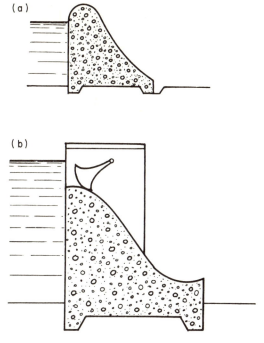

Fig. 15 Types of gravity dam: (a) normal gravity dam; (b) gravity dam with radial gate

resulting risk of damage to the downstream foot. This, therefore, has to be protected by a drainage system, comprising a drainage mat and a system of drains inside the dam itself.

With a dam of rockfill or bulky material, a water-tight core is incorporated into the structure, or a water-tight membrane can be fixed to the upstream face. The downstream foot must be protected by good drainage for all types of construction. The water-tight core can be of clay, but a sheet pile apron (see Fig. 16(c)) can also be installed in the centre of the dam. The upstream membrane can be a coating of clay, plastic sheeting, bitumen, etc.

An earth dam is a sensitive and vulnerable structure even if the fill has been compacted. Rapid emptying of a reservoir, as occurs when the lock-gates for an MHPS are opened, affects the upstream facing; the discharge from the opening attacks the downstream facing, and, finally, internal erosion damages the structure and causes cracks to appear.

Particular care has to be taken when designing spillways, especially if earth dams are used. If a spillway is incorrectly sized for the passage of a flood, the structure can be seriously damaged and may be destroyed. Again, construction has to be within the financial framework of the MHPS and so its size is selected to suit the chosen flood discharge. The fabric of the structure is monitored and measures are taken to prevent erosion. If the local relief

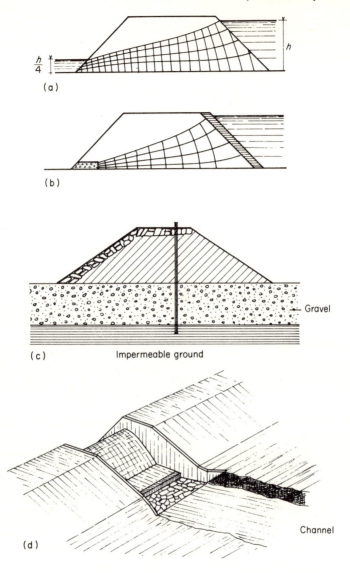

Fig. 16 (a) Earth dam showing flow pattern through homogenous materials. (b) Earth dam showing upstream membrane and downstream drainage. (c) Mixed material (earth and sheet pile) dam. (d) Central spillway

permits, an improvised spillway can be built away from the dam itself, but if this is not possible then the spillway is placed in the centre of the dam (a central spillway) or on one of the banks, when it is called a side-channel spillway.

Hill dams allow reservoirs of the order of 10 000 m^3 to be built and earth dams are installed in small valleys. They are several metres in height and constructed from earth excavated on the site and bulldozed into place. The loose material should not contain too much clay (30% clay + sediment, 50% fine sand, and 20% coarse sand), and the foundations should not settle or slip.

Micro hydro-power stations have been installed using such a construction but their output is low and very irregular.

(c) *A particular example: lock-gates on a river*

Locks (Fig. 17) were known to have been used in France in 1528 on the River Ourcq and in 1538 on the River Vilaine and the River Lot, but have in the main been developed over the last century. Only river locks can be used for the installation of MHPSs as low flows in canals are usually insufficient, although it may be possible in some cases to increase the flow to improve the efficiency of the turbines, as has been done by the EDF on the canal alongside the Garonne.

Under certain conditions, waterfalls at abandoned lock-gates can be used to produce hydroelectricity. In France, they can be as high as 4.18 m, and average about 2 m. The dam created by the lock-gate forms the intake and the MHPS can be placed inside the lock, to one side if the lock is still in use.

As with low head installations, the operation of such micro hydro-power stations is affected by floods, which can reduce the head or cancel it out through a rise in the downstream water-level.

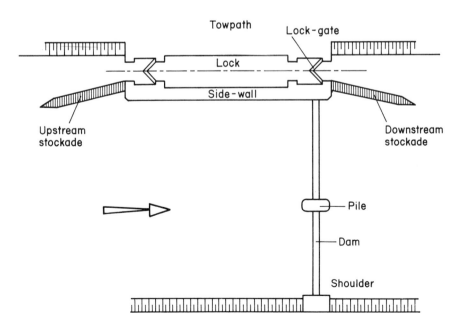

Fig. 17 Plan of a river lock

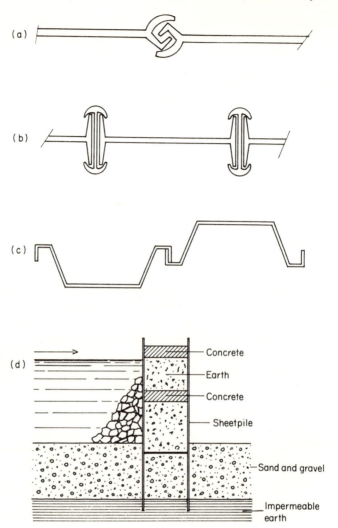

Fig. 18 (a) Lackawanna, (b) universal, and (c) Larsen sheet piles, together with (d)
mixed structures which use sheet piles

(d)   *Other types of intake*

Other types of simple structure can be used on torrents and rivers. However,
their cost must be determined in advance, as they involve the use of
specialized materials which are generally heavy and difficult to transport.
Foundations involving excavation, such as where caissons of reinforced
concrete sections are sunk into the ground as the earth is removed by a
mechanical digger, can form the support for the intake. Single or double sheet
pile shutters (Fig. 18) allow different types of low walls (< 3 m) to be built.

Although more difficult, construction on driven steel, concrete, or wooden piles can be employed for a poor site such as in a sedimentary or weathered area. The cost of the site works needed for installing sheet pile shutters depends on the availability of pile-drivers with rams suited to the required penetration.

Finally, it should be noted that rudimentary intakes can serve very well. Examples include submerged intakes across rivers, constructions of planks held in a vertical frame anchored in rockfill, and flexible gabion dams with a clay core (see Fig. 19).

3.4.3 Inlet and outlet channels

These channels fall into two types:

(a) Open channels, which are simply to direct the required flow of water to the forebay or full-pipe and from the turbine outlet. In the first two

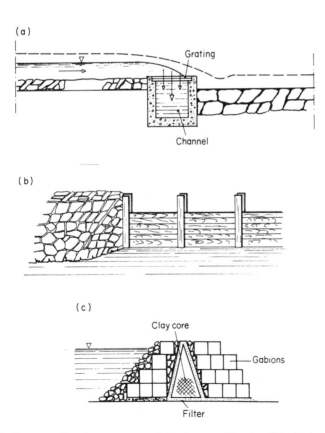

Fig. 19 Rudimentary intake structures: (a) submerged intake; (b) plank and rockfill; (c) gabions

cases, they are called *head-races*, and in the last case they are called
tail-races. The gradients of these channels are generally very low.

(b) Full-pipes, which convert the potential energy of the head into
pressure energy which is directly usable by the turbine, by means of a
closed pipe which is always full. In this case, the gradient can be very
steep.

In the open channels, there is a loss of head which ensures the maintenance
of the flow with an open surface. The head loss (J) can be shown to equal the
gradient (i) of the open surface. There are a number of simple equations
relating the rate of flow and the gradient (i), of which the most frequently
used is the Manning–Strickler equation:

$$u = kR_h^{2/3}\, i^{1/2},$$

where J is the head-loss ($J = i$), u is the speed in the channel, R_h is the
hydraulic radius ($=s/B$, i.e. the wet section to wet perimeter ratio), and k is
the Strickler roughness coefficient. (For polished sufaces $k = 100$; for a
concrete surface $k = 65$–85; for rough rock $k \simeq 30$.)

As an example, consider the values $R_h = 1$ m, $u = 2$ m s^{-1}, $Q = 12.5$ m^3
s^{-1}. Then

$$J = 4 \times 10^{-3} \text{ or 4 m km}^{-1} \text{ in rough rock,}$$

$$J = 4 \times 10^{-4} \text{ or 0.4 m km}^{-1} \text{ in polished steel.}$$

For MHPSs, discharges can range from 0.2 to 3 m^3 s^{-1}, and the cross-
section will be between 0.1 and 5 m^2 (for $u = 2$ m s^{-1}). These represent
rectangular sections of from 0.5 m × 0.2 m to 1.25 m × 4 m.

The channels should be sufficiently water-tight to prevent water from
escaping to such a degree that the discharge is reduced and the underlying
area becomes saturated, increasing the risk of slippage. The waterproof lining
can be achieved simply by facing the channel with clay, as used for the dam
core, or by using a cement coating which is resistant to vegetation, the effects
of drying and attack by burrowing animals. A bituminous coating may be
used, giving longer-lasting waterproofing, or a polyethylene coating, which
cannot be perforated by plants.

There are two types of head loss in the pipes: distributed head losses which
are proportional to the length of the flow being considered, and localized
head losses which are related to changes in the pipe cross-section, bends, etc.

(a) *Sizing*

The calculation of the pipe dimensions involves determining the economic
cross-section, S_e. The average speed in the pipe is $u = Q/S_e$, which is

generally between 3 and 8 m s^{-1} for a total head loss of 2–7%. A second strength of materials calculation allows the wall thickness to be determined from the forces acting on it. In addition to these constraints, some very important phenomena must also be allowed for, such as oxidation and fatigue caused by the cyclical effects of some forces.

Steel pipes are most frequently used. They are usually pre-formed tubes of rolled and welded steel, which are welded together on the site. PVC pipes can be used for small MHPSs (where $P < 100$ kW and h 80 $<$ m). These are not ideal as they are very fragile and can be damaged by falling stones. Additionally, they deteriorate under the effects of ultraviolet radiation, to become even more brittle. This effect can be avoided by burying them or by using a protective covering such as tarred canvas. Polyethylene tubes can also be used as they are less fragile and are not aged by sunlight. PVC or polyethylene tubes have a low limiting pressure σ_a, of about 100 MPa.

3.5 THE ROLE OF THE ENGINEER

3.5.1 Design of an MHPS and management of the construction

The contracting authority usually assigns the engineering studies to a design consultancy bureau and a master of works.

The design of an MHPS is carried out in several phases – the feasibility study and the brief draft proposal, application for a permit (or licence), detailed proposal, and supervision of the construction.

(a) First phase: Feasibility study

The information collected during this phase is to establish the viability of the site and to allow a decision to be made as to whether the project should be continued.

From the hydrological, geological and geotechnical data for the site, a schedule of equipment is drawn up and feasibility studies are established for each possible solution. The reconnaissance work on the site and the laboratory experiments can be carried out during this phase.

These initial studies define the capital cost and allow the viability of the project to be evaluated in terms of the anticipated income and the outgoings for interest charges, development costs and the costs of replacement and maintenance of the equipment.

(b) Second phase: Application for a permit or licence

This is discussed in detail in Chapter 6.

Summary 7 Phases for installing a micro hydroelectric power station

Phase	1	2	3	4	5	6	7	8	9	10	11	12	13	14	15	16	17	18
First evaluation: decision to proceed with studies	▬																	
Reconnaissance studies		▬																
Study of environmental impact			▬															
Preparing and filing the application for a licence			▬															
Examination of the application by the administrating authorities				▬▬▬														
Supplementary examination of site; preparation of specifications and invitation to tender				▬▬▬▬														
Selection of contractor								▬										
Order electromechanical equipment; civil engineering work on intakes, pipes, buildings, etc.; erecting cables for connection to the network										▬▬▬▬▬								
Installing electromechanical equipment Testing and commissioning of the MHPS															▬▬▬			

(c) Third phase: Detailed proposal

The detailed proposal may include supplementary information on the heads, the discharge and the hydrological pattern of the river. Reconnaissance work can be carried out during this phase. The detailed proposal generally allows requests for tender to be made to the construction companies, after which the company is selected on the basis of recognized technical ability and tender price.

The final work programme is usually drawn up by the construction company with time schedules and detailed technical specifications.

(d) Fourth phase: Supervision of the construction

The master of works can be involved in different ways. He can act as technical assistant to the contracting authority or he can be in charge of the overall supervision of the construction. In this case he is responsible for overall coordination, deliveries, and the testing and commissioning of the system.

3.5.2 Cost of the engineering aspects

Cost evaluation studies carried out for local authorities are usually tightly controlled. In France, for example, public control of such assessments is covered by legislation which includes a statement of the responsibilities of the engineer. Every public contract is subject to the laws which apply to the public sector in general. Under these, the consulting engineer must adhere to the objectives defined by the contracting authroity as to cost, time-scale, and quality control throughout all phases of project planning and construction.

In general, about 7–15% of the total capital cost is devoted to the feasibility studies, drawing up of applications for permits, tenders, selection of contractor, and supervision and monitoring the construction. By way of guidance, it should be noted that if a plant power of less than 400 kW is to be viable, its total cost must be less than 1 FF (1980) per kilowatt-hour produced.

BIBLIOGRAPHY AND FURTHER READING

ANONYMOUS (1969). *La reconnaissance des sols*, Laboratoire des Ponts et Chaussées, Paris.

ANONYMOUS (1970). Technique d'approche pour la connaissance des terrains à urbaniser. *Les Cahiers du Beture*, February 1970.

ANONYMOUS (1976). Les plans des zones exposées aux avalanches. *Technical information sheet no. 24*, Centre technique Génie rural Eaux et Forêts, Antony.

ANTOINE, P. and BARBIER, R. (1973). Les problèmes posés par les barrages de faible hauteur. In *Conférence au comité français de géologie de l'ingénieur, Annls Inst. tech. Bât. Trav.*, Suppl. 312, 25–60.

ANTOINE, P. and FABRE, D. (1980). *Géologie appliquée au génie civil*, Masson, Paris.

ASTE, J. P. DESURMONT, M. and SIMON, J. M. (1978). Confortement de versants naturels, objet et moyens. *Internal report no. 78 SGN 025 GTC*, BRGM, Orléans.

BAGNÈRES, M. (1980). Conception des microcentrales. Paper presented at ALPEX-PO, 4e Salon intenationale de la Montagne, Grenoble.

CAILLAT, P. and CAILLET, B. (1972). *Connaître et prévenir les avalanches*, Albin Michel, Paris.

COMMISSION INTERNATIONALE DES GRANDS BARRAGES (1973). *Leçons tirées des accidents de barrages*, CIGB, Paris.

DUFFAUT, P. (1980). Génie civil pour microcentrales hydroélectriques. Unpublished paper, BRGM, Orléans.

GALABRU, P. (1963). *Les fondations et les souterrains*, Eyrolles, Paris.

GOGUEL, J. (1959). *Application de la géologie aux travaux de l'ingénieur*, Masson, Paris.

GOGUEL, J. (1980). *Géologie de l'environnement*, Masson, Paris.

HUMBERT, M. (1972). Les mouvements de terrains. Principes de réalisation d'une carte prévisionnelle dans les Alpes. *Bull. BRGM, Section III*, (1), 13–28.

HUMBERT, M. (1980). Les risques géologiques dans la législation actuelle. Bilan et perspectives, *Internal report no. 80 SGN 584 GEG*, BRGM, Orléans.

LETOURNEUR, J. and MICHEL, R. (1971). *Géologie du génie civil*, A. Colin, Paris.

L'HERMITE, R. (1962). *Au pied du mur*, Eyrolles, Paris.

MAYER, A. (1959). *Précis de mécanique des sols*, A. Colin, Paris.

MENCL, V. and Zaruba, Q. (1976). *Engineering Geology*, Elsevier, Amsterdam.

MINISTÈRE DE L'AGRICULTURE . *Retenues collinaires*, La Documentation française, Paris.

MINISTÈRE DE L'AGRICULTURE (1977). *Technique des barrages en aménagement rural*, La Documentation française, Paris.

RANDET, P. (1970). Les moyens contre les avalanches. *Le Moniteur*, 12 December 1970.

TER MINASSIAN (1977). Le barrage sur mesure. Actuel développement no. 21.

CHAPTER 4

Electromechanical equipment

4.1 HISTORICAL BACKGROUND

The idea of transforming the kinetic and potential energy of water in a river into useable mechanical energy is very old, and the first machines for this transformation were waterwheels which could be used on small heads.

The first waterwheels had vertical axes, and are called Norse, horizontal, Greek, Pyrenean or Arabic wheels. This type of waterwheel is described in documents dating from the 15th century, and is also mentioned in an 8th century Irish manuscript. It was widely used throughout Europe with the exception of England and Wales. An improved wheel, called a tub-wheel,

Fig. 20   A tub-wheel at the Moulin de Norois (Cantal)

was typical of Occitania in southern France, the Pyrenees, and Spain. Figure 20 illustrates a wooden wheel of this type which was in operation in the Cantal until the 1930s. Similar wheels, but made of metal, probably dating from the beginning of this century, are still in operation in the upper valley of the Allier.

Bernard Forest de Bélidor (1693–1761), a military hydraulic engineer, described such wheels with concave paddles in his monumental work *L'architecture hydraulique* published between 1737 and 1753. These particular wheels were installed at Le Basacle on the Garonne.

Although many factors favoured tub-wheels, in particular their ease of construction and installation and their dependable and convenient operation, their efficiency was rather low, reaching 20% at best but often not exceeding 15%.

Besides this vertical axis wheel, slow hydraulic wheels with horizontal axes and much better efficiencies were also developed. Thus, the Roman or 'Vitruvian' wheel, attributed to the 1st century BC Roman engineer and architect, Marcus Vitruvius Pollio, has evolved into three classical types – the overshot or bucket-wheel which receives the water on its upper part, an intermediate type known as the breast-wheel, and the undershot or paddle-wheel (Fig. 21).

After a series of well designed and careful experiments, the English engineer and manufacturer John Smeaton (1724–92) showed that overshot wheels had an efficiency of at least 60%, or twice that of undershot wheels, which resulted in an eventual movement away from undershot wheels in the British Isles.

In France, where research on hydraulics had begun to diversify, engineers established two fundamental principles for an ideal hydraulic engine. The first is that the water must always enter without producing turbulence. The second states that as it passes through the engine, the water must lose the initial velocity that it had on entry. These two conditions must be satisfied if there is to be no energy wasted either as turbulence or as residual kinetic energy. By applying these two principles, the French engineer Jean-Victor Poncelet (1788–1867) fitted an undershot wheel with a sufficient number of blades of the correct shape and increased the efficiency to 60–70%. As a consequence, the use of this design spread to many regions.

Using this type of design, tidal wheels, the forerunners of the tidal power station, were constructed in France and England up to the 17th century. They were designed to use the energy of the currents in both directions and to operate for about 16 h per day. The incoming tide fed a reservoir which stored the energy for use when the tide was ebbing.

Nevertheless, the limited knowledge of hydraulics meant that these machines remained rudimentary and had a low power output, as most of the head was dissipated in the feed through an open conduit.

During the 18th century, Daniel Bernoulli (1700–82) and Léonhard Euler (1707–83) developed the fundamentals of hydrodynamics, and thus prepared

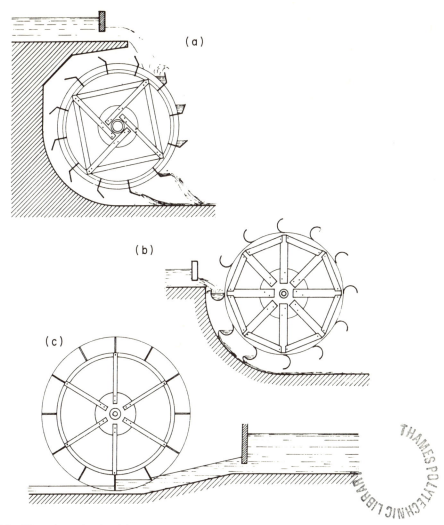

Fig. 21 Vitruvian wheels: (a) overshot wheet; (b) breast wheel; (c) undershot or paddle-wheel

the way for the advent of modern hydraulic machines. Water power had come of age, and in the 19th century 'turbines' set the standards of performance.

Among other things, Euler developed the theory of the reaction turbine, which in principle is that of the hydraulic jet. Jets of water are released at the curved extremities of two or more arms. The water is admitted to the centre of the engine through a water-tight coupling, and the arms rotate because of the reaction effect. Jean-Charles de Mannoury d'Ectot (1777–1822) designed an engine of this type, which was called a 'hydraulic wedge' and which operated forges and engineering workshops in Normandy.

Noting that water always entered and left on the same side of an undershot wheel, Poncelet observed that under these conditions it was impossible to force all the water to come into contact with the wheel as it travels over it. This was not an acceptable state of affairs. He proposed a method of resolving this difficulty by laying the wheel flat, which allowed the water to enter over the complete circumference and to reject it through the centre at zero velocity. Thus, the radial centripetal flow machine with total injection was invented. In 1822, the military engineer Claude Burdin (1790–1873) submitted a paper to the French Academy of Sciences entitled 'Hydraulic turbines or high-speed rotating engines', in which he described the engine which was to be built and tested between 1823 and 1827 by his pupil Benoit Fourneyron (1802–67). This paper used the word 'turbine' for the first time (formed by Burdin from the Latin word *turbo*, meaning a spindle wheel).

Fourneyron went on to improve this radial centrifugal total injection wheel, initially with free flow, and then with forced flow by means of a diffuser whose shape resembled that of today's spiral casing. Over a period of 30 years he designed and constructed more than 100 turbines throughout Europe, with some being exported to the USA.

Meanwhile, one limitation of these machines had been noticed. They could operate well only within very precise flow limits, and their efficiency fell significantly outside this range. As a consequence, after 1837, Jonval or Fontaine turbines, which were of the axial type with controllable partial injection, rapidly became a significant competitor.

During the same period, engineers in the USA tended to concentrate on centripetal-type machines, whose development was perfected by James Francis (1815–92).

Turbines with partial injection were devised to exploit high heads and low flows. Escher-Wyss in 1840 introduced the Zuppinger tangential turbine with a rectangular section jet. The principle of these machines, based on that of the Poncelet wheel, ultimately led to the bucket wheel and to the cylindrical jet, perfected by a group of Californian engineers, one of whom was Lester Pelton (1829–1908). Initially, the flow was controlled by means of a gate and not by the needle valve used in modern turbines.

In response to the same preoccupations, the French engineer Girard launched a highly successful cetrifugal radial turbine with partial injection in 1851. Donat Banki, Professor of Hydraulics at the University of Budapest, also designed a turbine of the same type with a double passage of the fluid through the wheel. This machine was finally perfected at the beginning of this century by the English engineer A. G. Mitchell.

Finally, comparatively recently (i.e. in 1924), the need to obtain high specific velocities for use with small heads and large flows and a minimum number of units led to the development of propeller turbines such as that perfected by Victor Kaplan (1876–1934).

As these devices evolved, ever-increasing capacities were being harnessed. For example, the Girard or Jonval turbines, regarded as the most powerful in

1900, developed 1500 hp (1100 kW), whereas today powers of the order of 200 000 kW can be attained.

Since the appearance of full-pipes, harnessed heads have increased dramatically, as has the flow of water used. Because of these improvements, priority was given to those sites with the largest energy potential, and small streams that could only supply the immediate locality were abandoned. This progression in performance is shown in the diagrams in Fig. 22.

Initially, variations in flow through the machines were not important, but soon the requirement for obtaining constant rotation speeds, firstly for machinery such as weaving looms, then for the production of alternating current at a constant frequency, led to the introduction of control systems which are often complicated and expensive.

Finally, two significant developments should be described which considerably improved turbine performance. The first was the introduction of adjustable guide vanes which, whatever the flow, allow turbulence-free operation at the entry of the wheel. The second was the linkage with an alternator which allowed the efficiency to rise to 90% in comparison with the 65–70% of mechanical systems with pulleys and belts.

Thus, as a result of the prodigious advances in technology over the last two centuries, hydro-power, which was created for local use in a watermill, has changed its emphasis towards ever-increasing power production. Distribution of the energy has also spread out in an ever-widening circle as electricity has become a major energy carrier. The latest energy policy conditions and technological improvements now allow us to close the circle by rediscovering a mode of hydroelectricity production which is suited to local requirements and with much lower power ranges than those of the gigantic central generating authority power stations.

4.2 SOME GENERAL POINTS ON HYDRAULIC TURBINES

A hydraulic turbine is a rotating machine, basically consisting of a paddle-wheel, to capture the kinetic and pressure energy of a fluid and transform it into the directly usable mechanical energy of a rotating shaft. The turbine can be either submerged in the fluid in a water chamber, or placed at the end of a full-pipe.

4.2.1 Gross, net and effective heads

Consider a hydraulic circuit with a turbine T as shown in Fig. 23. The potential energy between A and B is transformed into mechanical energy in the turbine and into heat through head losses in the pipe.

Using Bernoulli's theorem between A and B, we obtain

$$h_g = z_A - z_B = h_n + \Delta h,$$

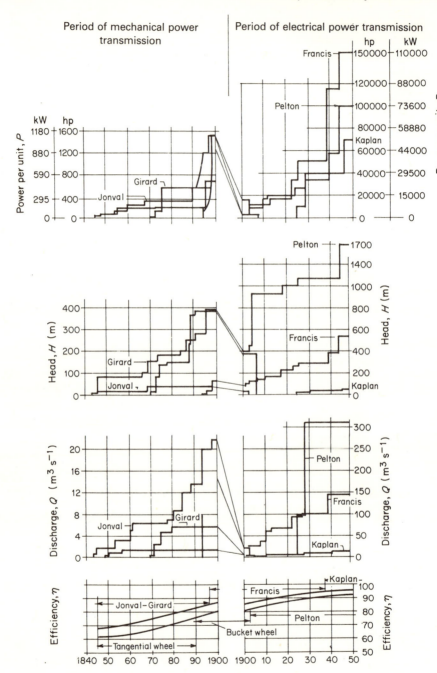

Fig. 22 Development of hydraulic turbines between 1840 and 1950
(from Vivier, 1964)

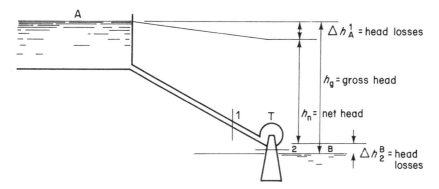

Fig. 23 Hydraulic circuit with turbine

where h_g is the gross height, Δh is the sum of the head losses in the pipe between A and 1 and between 2 and B, and h_n, the net height, represents the energy supplied to the turbine per unit weight of the fluid. In this case, h_n is the useful head loss serving to operate the turbine.

The net power supplied to the turbine is therefore

$$P_n = h_n \bar{\omega} q_v = h_n g q, \tag{1}$$

where $\bar{\omega}$ is the weight density of the fluid $= \rho g$, q_v is the volumetric discharge, g is the acceleration of gravity, q is the mass discharge, and ρ is the mass density of the fluid. The standard definitions of powers and heads are given in Chapter 6.

For the turbine illustrated in Fig. 24, the indices e and s indicate the entry and exit of the machine. The net head is $h_n = h_e - h_s$, and represents the

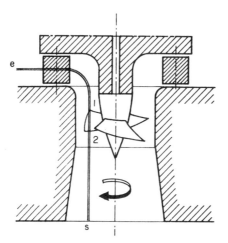

Fig. 24 Head losses in a turbine

energy available to the engine (Fig. 24). Δh_e^s represents the head losses through friction of fluid on fluid and fluid on the pipe walls, or through turbulence at the entry to the vanes.

The effective head, h_E, defined by

$$h_n = h_E + \Delta h_e^s, \tag{2}$$

represents the energy actually transferred to the wheel.

4.2.2   General layout

A hydraulic turbine comprises three elements:

  (i)   The *runner*, which transforms hydraulic energy into mechanical energy. It has buckets (for example, the Pelton wheel), paddles, or blades, either in the air (as in the Banki–Mitchell turbine) or in pressurized pipes in reaction turbines (such as in Francis, propeller, and Kaplan turbines). The buckets or paddles in the air change the velocity of the fluid while the blades under pressure change the velocity and the pressure of the fluid.
  (ii) The *distributor*, which gives the water the velocity needed so that it reaches the wheel under conditions which minimize losses and thus transfer as much as possible of the pressure and kinetic energy.
  (iii) The *draught-pipe* (in reaction turbines), which is included to recover as pressure energy the residual kinetic and potential energy of the water at the exit of the wheel and to drain it off downstream.

(a)   *Classification according to the motion of the fluid relative to the wheel*

In a radial turbine such as the Banki–Mitchell type, the fluid motion is along the radius of the wheel (Fig. 25(a)).

If the fluid motion is along the axis of the machine, it is an axial turbine, as with Kaplan and propeller units (Fig. 25(b)).

The motion can be a combination of these two, in which case it is a mixed type of turbine, such as the Francis machine (Fig. 25(c)).

In certain cases, such as in the Pelton wheel (Fig. 25(d)), the motion of the fluid can be tangential to the wheel. The wheel is then of the tangential type.

A further classification can be established according to whether the wheel uses solely the kinetic energy of the fluid (impulse turbines, e.g. Pelton and Banki turbines) or simultaneously uses both the kinetic and pressure energies (as in reaction turbines such as propeller, Kaplan and Francis machines).

(b)   *Some other definitions*

A partial injection turbine is one in which the fluid is only injected over a fraction of the perimeter of the runner (e.g. Pelton and Banki–Mitchell

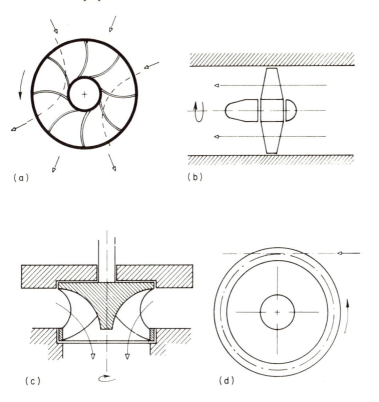

Fig. 25 Classification of turbines based on the flow of fluid through the wheel: (a) radial turbine (Banki–Mitchell); (b) axial turbine (propeller, Kaplan); (c) mixed turbine (Francis); (d) Tangential turbine (Pelton)

wheels). In contrast, a total injection turbine is one in which the fluid is injected over the whole perimeter of the wheel as in Kaplan, propeller, and Francis units.

A vertical axis turbine is one whose axis of rotation is vertical and which therefore rotates in a horizontal plane (with the exception of the Banki–Mitchell turbine, the others can be either horizontal or vertical).

A multiple wheel turbine is formed of two or three wheels rotating on the same shaft (e.g. Pelton or Francis machines).

The degree of opening of a turbine is the ratio of the discharge q to the maximum usable discharge q_{max}.

4.3 ELEMENTS OF TURBINE OPERATION

In a micro hydroelectric power station, the turbine is the primary element for the transfer of energy.

As it is a dynamic component, the efficiency of the machine can be defined in terms of kinetic factors (e.g. the triangle of velocities) and energy factors (the calculation of torque and energy losses).

Finally, determination of the geometrical characteristics of a particular machine is derived from a definition based on its similarity to an ideal machine (see Section 4.4).

4.3.1 Velocity diagram

The absolute velocity vector **C** at a point M of the flow is a tangent to the trajectory at M, and can be decomposed into a helical system (see Fig. 26) consisting of a rotating velocity C_u perpendicular to a radius of the runner, an

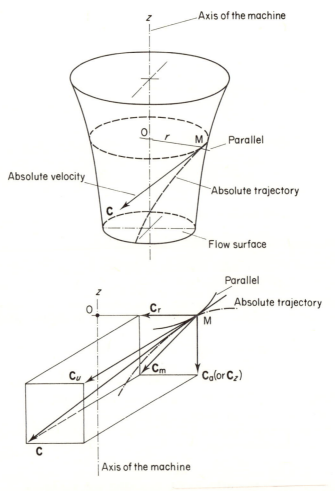

Fig. 26 Velocity of fluid in the turbine runner. C_u determines the exchange of energy between the fluid and the runner; $C_m = C_a + C_r$, the longitudinal component, determines the turbined flow. (C_u and C_m are in a plane tangential to the flow surface.)

axial velocity C_a (or C_z) parallel to the axis of the runner and a radial velocity C_r along a radius.

In the active zone of the runner, a frame of reference is often used which rotates with respect to some absolute frame such as the casing of the turbine, at a rate ω equal to the velocity of the rotation of the wheel.

The velocity at the point M relative to the wheel is therefore

$$\mathbf{u} = \omega \times \overrightarrow{OM} \, ,$$

where $\overrightarrow{OM}$ is the polar radius, and $u = \omega r$ is the feed rate. A relative rate of flow $\mathbf{W}$ can now be defined for the fluid with reference to the runner such that (Fig. 27)

$$\mathbf{C} = \mathbf{u} + \mathbf{W}. \tag{3}$$

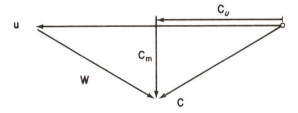

Fig. 27   Triangle of velocities

4.3.2   Calculation of torque on the shaft – indicated power

The moment (M) about the z-axis due to the action of the fluid on the walls of a channel is ΔM_z (Fig. 28).

Taking moments about the z-axis,

$$\Delta q(r_2 C_{u2} - r_1 C_{u1}) = -\Delta M_z + Mt/z$$
(forces due to pressure and the weight of the fluid)

By summing over all of the channels,

$$\Sigma \Delta M_z = C_E = \Sigma \Delta q(r_1 C_{u1} - r_2 C_{u2}).$$

(By symmetry, $\Sigma Mt/z = 0$.) Hence, the gross torque is

$$C_E = q(r_1 C_{u1} - r_2 C_{u2}) \tag{4}$$

with q = total mass flow.

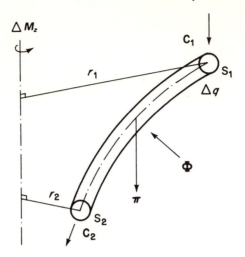

Fig. 28 Action of fluid on the sides of the wheel: π is the weight of fluid; Φ is the blade reaction; C_1 is the entry velocity; C_2 is the outlet velocity. Thus

$$\Delta M_z = \Delta q(r_2 C_{u2} - r_1 C_{u1}) + M\pi_z$$

$$M\pi_z = \text{pressure forces} + \text{fluid weight}.$$

As the rate of rotation of the wheel is ω, the indicated power is

$$P_E = C_E \omega,$$

whence

$$P_E = q(u_1 C_{u1} - u_2 C_{u2}) \tag{5}$$

since $\omega r_1 = u_1$ and $\omega r_2 = u_2$.

4.3.3 Operation of impulse and reaction turbines

At the inlet to the distributor of a turbine, the energy is presented partly in a kinetic form $C^2/2g$ and partly in a pressure form $p/\bar{\omega}$. The distributor then acts to transform the pressure energy $p/\bar{\omega}$ partially or completely into kinetic energy (speed C_0).

(a) *Impulse turbine*

The available energy at the inlet to the distributor is transformed entirely into kinetic energy. As a consequence, the water leaves this component at atmospheric pressure $p_0 = p_{atm}$ in the form of a free water stream. These are the Pelton and Banki–Mitchell type turbines.

(b) Reaction turbines

The energy of the water at the outlet of the distributor is presented partly as kinetic energy and partly as pressure energy. The water enters the runner at a pressure p_0 greater than atmospheric pressure p_{atm}, and during its passage through the turbine it undergoes an expansion from p_0 to p_1. Machines of this type include Francis, propeller and Kaplan turbines.

(c) Degree of reaction

The degree of reaction is the ratio of the pressure energy that the water has at the entry to the runner to the total energy corresponding to the usable head. It can be expressed as a function of pressures p_0 and p_1 by the equation

$$\sigma = (p_0 - p_1) / \bar{\omega} h_n.$$

If $p_0 = p_1$, $\sigma = 0$, and it is an impulse turbine (e.g. Pelton and Banki turbines).

If $p_0 \neq p_1$, $\sigma \neq 0$, and it is an impulse turbine (e.g. Francis, Kaplan and propeller turbines).

4.3.4 Energy losses

There are two types of energy loss in hydraulic turbines, internal or pressure losses, and external losses. These are very different in character.

(a) Internal or pressure losses

These are caused by the flow of water in the distributor, the runner, and the draught-pipe. They are purely hydraulic in nature and appear as a loss of pressure. They include the following:

(a) Losses through friction between the layers of water and between the water and the containment walls. These losses are proportional to the square of the velocity (and thus of the discharge).
(b) Losses through turbulence in the water at the runner inlet. These arise because the relative velocity **W** is not tangential to the vane. They cannot normally be avoided under part-load operation which is outside the optimum regime for which the turbine was designed.

(b) External losses

These arise through imperfections in the construction of the system. They are expressed not as head losses, but as power losses or as leakage flow. Mechanical losses include the friction in the various bearings, shafts or pressure seals. They also include the internal friction losses in the runner in

Francis turbines and, for automatically controlled turbines, the energy absorbed by the control components. Volumetric losses or losses through leaks at joints are related to the inevitable leaks which exist between the fixed components of the machine and the moving runner. Residual velocity losses are represented by the residual kinetic energy of the water at the outlet of the draught-pipe. These are deduced from the heads outside the installation. Cavitation losses, as will be shown later, result from vaporization of the water under certain conditions of under-pressure and can lead to noisy operation of the machine because of the resultant vibration. They appear as a decrease in efficiency and an increase in the erosion on the active surfaces of the machine.

4.3.5 Efficiencies

For each of the above losses there is a corresponding efficiency.

(a) *Pressure (or hydraulic) efficiency (η_h)*

This relates to the pressure losses through friction and turbulence, and corresponds to

$$\eta_h = h_E/h_n = \frac{\text{energy effectively used by the turbine}}{\text{energy available to the turbine}}$$

$$= (u_1 C_{u1} - u_2 C_{u2})/gh_n.$$

The pressure efficiency of the installation can also be defined as

$$\eta_h \text{ (installation)} = \frac{h_E}{h_g} \left(\frac{\text{effective head}}{\text{gross head}} \right)$$

(b) *Volumetric efficiency (η_v)*

A fraction of the flow, called the leakage flow, q_f, bypasses the runner through the water-tight joints. A volumetric efficiency can be defined as

$$\eta_v = q/(q + q_f),$$

where q is the useful flow.

(c) *Mechanical efficiency (η_m)*

This is related to the various losses through mechanical friction. By convention, the mechanical power available at the shaft is defined from $h_a = P_a/qg$, where h_a is the mechanical head.
   Hence,

$$\eta_m = p_a/P_E = h_a/h_E.$$

(d) *Overall efficiency* (η_g)

This is the ratio of the mechanical power, P_a, available at the shaft to the net power, P_n, available to the turbine. It is the product of the three partial efficiencies listed above:

$$\eta_g = \eta_h \eta_v \eta_m = P_a/P_n = P_a/((q + q_f)gh_n). \qquad (6)$$

(e) *Efficiency curves*

The overall efficiency of a turbine varies with the degree of opening (see Section 2.2) because of the losses through turbulence at the inlet to the runner. Because of this, the specification of a turbine generally includes a curve showing the variation of efficiency with the degree of opening. In Fig. 29, these efficiency curves are given as a function of their rated speeds (Ω) for the principal types of turbine.

It should be noted that impulse turbines such as the Pelton and Banki types are fairly insensitive to the degree of opening, as are Kaplan turbines whose vanes can be adjusted to variations in W_1 (the relative velocity of the water).

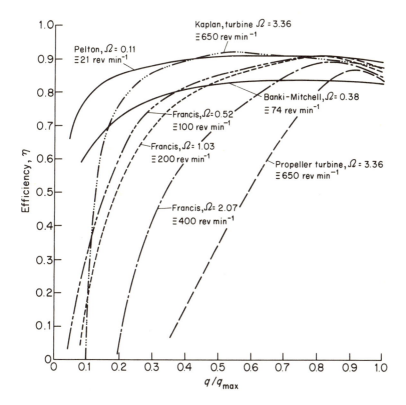

Fig. 29 Variation of efficiency with degree of opening for different types of turbine

In contrast, Francis and propeller turbines have efficiency curves which become more sharply peaked as the specific velocity increases since the relative speed W is greater, and hence ΔW is also greater when far from the optimum regime.

As will be shown in Section 4.4.2, the overall efficiency of the turbine (η_g) does not obey simple scaling laws. There is a scale effect which dictates that the efficiency of a turbine increases as its power, and hence size, increases. This effect operates against micro hydro-power stations.

4.4   SCALING LAWS FOR HYDRAULIC TURBINES

4.4.1   Specific velocity

If two geometrically similar machines S and S′ with geometrically similar flows are considered, then dimensional analysis leads to the definition of a scale factor,

$$\alpha = (P_n'/P_n)^{1/2}\,(h_n'/h_n)^{-3/4},$$

and a dimensionless product,

$$C_k = n(P_n/\rho)^{1/2}\,(gh_n)^{-5/4},$$

where $n = 60\omega/2\pi$ is the speed of rotation of the wheel in revolutions per minute.

The specific velocity n_s of a turbine S is defined as the rotation speed of a geometrically similar turbine operating dynamically to scale which produces a power of a 1 hp under a net head of 1 m with the same optimum pressure efficiency as that of the turbine S:

$$n_s = n(P_n^{1/2}/h_n^{5/4}), \tag{7}$$

where n_s is expressed in revolutions per minute, P_n in horsepower, and h_n in metres.

Similarly, one can define

$$\Omega = \omega(P_n/\rho)^{1/2}\,(gh_n)^{-5/4},$$

called the angular velocity coefficient, and

$$\Lambda = R\,\frac{(gh_n)^{1/4}}{(q_v)^{1/2}},$$

called the radius coefficient.

One can also define a specific runner diameter d_s from h_n and P_n such that

$$d_s = d(h_n^{3/4}/P_n^{1/2})$$

For water, $\rho = 1000$ kg m^{-3}, and the following practical equations apply:

$$\Omega = n_s/193.5 \quad \text{and} \quad \Lambda = 3.23 d_s.$$

4.4.2 Application to the classification of turbines

From experience, a hydraulic turbine is now characterized by its maximum efficiency condition, or its rated operation. Also, in order to permit comparison of models of different sizes under varying heads, it is usual to report the results per unit of head (1 m) and per unit of diameter of the runner (1 m). The power is then conventionally designated as P_{11} (in horsepower) and the rotation speed as N_{11} (in revolutions per minute). Figure 30 gives a graph

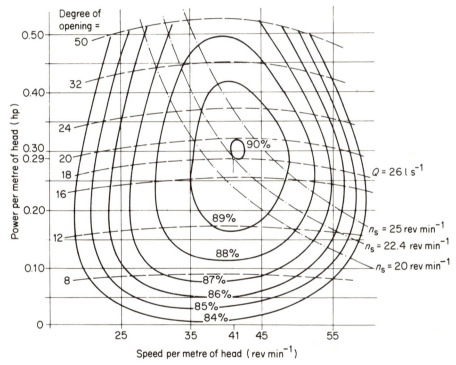

Fig. 30 Efficiency chart for a prototype Pelton turbine with one wheel and two jets.
— — — : curves showing the degree of opening of the injector;
— · — · — : curves showing the specific speeds.
The scaling equations are given by

$$N' = N\lambda \, (H'/H)^{1/2}, \qquad P' = \frac{P}{\lambda^2} \, (H'/H)^{3/2}, \qquad Q' = \frac{Q}{\lambda^2} \, (H'/H)^{1/2},$$

where λ is the scaling factor. Specific values are as follows:

Height	H'	= 1 m	$H = 1112$ m
Diameter	D'	= 1 m	$D = 3.1$ m
Rotation speed	N'	= 41 rev min^{-1}	$N = 428.6$ rev min^{-1}
Power	P'	= 0.29 hp	$P = 113\,000$ hp
Flow rate	Q'	= 26 l s^{-1}	$Q = 8500$ l s^{-1}

established for a Pelton turbine with two jets and one wheel of 1 m diameter. Also included are the isoefficiency curves, the values for the aperture of the injectors and the values for the specific velocities n_s. The coefficients for cavitation and other elements can also appear on the complete diagram for a reaction turbine (e.g. Francis, propeller or Kaplan turbines).

In this example, the rated operation is taken as a flow of 26 l s^{-1}, a specific velocity of 22.4 rev min^{-1}, a power of 0.29 hp (213W) and an efficiency of 90%.

The value of n_s (or Ω) uniquely characterizes a turbo-machine at this point and the application ranges of the various types of hydraulic turbines can be derived from this value.

For a machine with multiple wheels, the coefficient calculated for one stage is used.

By referring to the various definitions of the efficiencies in Section 3.5, it can be shown that only the pressure efficiency η_h obeys the scaling laws.

The overall efficiency of a hydraulic turbine does not therefore obey the scaling laws, and, in order to proceed from a test wheel to an actual wheel, the values of η_{11} (the efficiency of the model turbine) have to be corrected. The many formulae used by turbine engineers to do this all show that the correction factor, called the scale effect, decreases as the power generated by the machine increases.

4.5   VARIOUS TYPES OF TURBINE

4.5.1   Impulse turbines

(a)   *Pelton turbine: Description*

A Pelton turbine is formed of a wheel with a series of buckets with a median notch, in the shape of a double spoon. The water is sent under pressure through distributor nozzles, of which there can be between one and six lying in the plane of the wheel. In practice, there are never more than two in a micro hydro-power station. Because of the notch, the jets can strike two or three buckets. Each distributor is formed of an injector with a feed-pipe and is controllable to allow a full, partial or zero feed. A deflector completes the installation by deflecting the jet to avoid racing. A schematic diagram of a Pelton installation is shown in Fig. 31.

Range of application   Given the range of low specific velocities that it covers, this type of turbine is very suitable for large heads (> 200 m) and for low flows. In principle, with their high rotation rates (of the order of 500–1500 rev min^{-1}), these machines have good efficiencies over a wide range of variation in the degree of opening q/q_{max}.

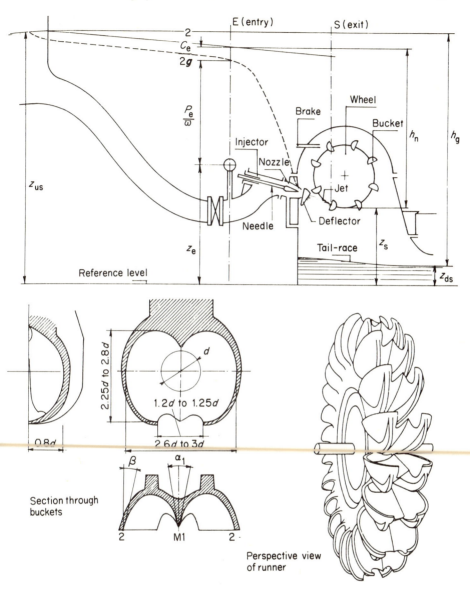

E (entry)   S (exit)

Wheel

Brake

Bucket h_n h_g

Injector

Nozzle

Needle

Deflector

Jet

Tail-race z_s

z_{us}

$\frac{P_e}{\varpi}$

z_e

Reference level z_{ds}

2.25d to 2.8d

d

1.2d to 1.25d

2.6d to 3d

0.8d

β α_1

Section through
buckets

2   M1   2

Perspective view
of runner

Fig. 31   A Pelton turbine

(b)   *Banki–Mitchell turbine: Description*

The Banki–Mitchell turbine is an adaptation of Vitruvian breast-fed wheels. As shown in Fig. 32(a), it is formed of a rectangular injector section, which can be adjusted by a profiled vane, and a double-impulse partial-injection wheel. The water acts first upon the cylindrical wheel in a centripetal manner,

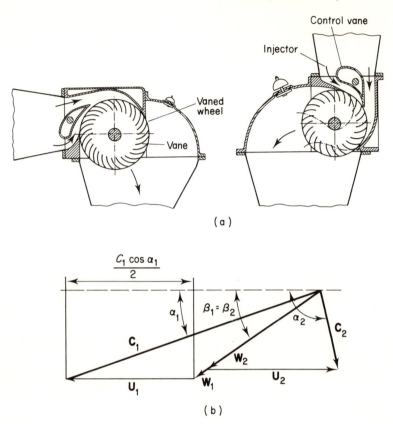

(a)

(b)

and then again in a centrifugal manner after it has passed through the interior of the wheel. The wheel is formed of two end-discs connected to each other by the ring of blades.

Range of application   Due to its capacity for adaptation to suit a wide range of patterns of operation, this type of turbine can be used for heads ranging from 1 or 2 m up to 200 m and for flows varying from 20 to 10 000 l s^{-1}. The rotation rates are also highly adaptable as they can vary from 50 to 2000 rev min^{-1}. However, as the ring of blades becomes more fragile with increasing power, due to the physical size of the machine, they have always been limited to powers less than 1000 kW.

Because of the properties given earlier, the efficiency can remain very stable ($\simeq 0.82$) from an opening of 0.08 to full rated flow. An elegant manufacturing technique permits the active part of the wheel to be restricted to suit the available flow, as shown in Fig. 32(d).

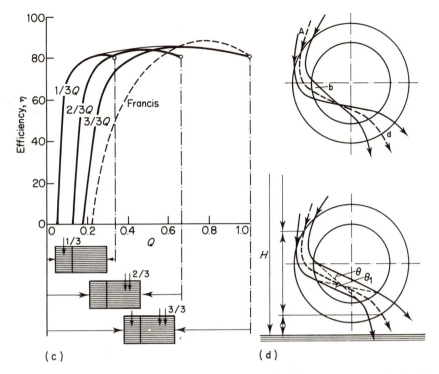

Fig. 32 A Banki–Mitchell turbine (after Ossberger): (a) arrangement of a Banki–Mitchell turbine; (b) triangle of velocities for a single vane; (c) variation of efficiency with injection rate; (d) interference of fluid streams through the wheel

This type of turbine, because of its compactness, its simplicity of operation and therefore of maintenance, its great flexibility in adapting to changes in the pattern of operation, and its simplicity of installation, is the ideal choice for many micro hydro-power stations.

4.5.2 Reaction turbines

This family of machines includes Francis or propeller-centripetal turbines, propeller turbines, and axial-type Kaplan turbines, which can be distinguished by the shape of their runners. As they are used for low to average heads, these machines have the same fixed components and control system.

(a) *Feed system and distributor*

For heads of less than 4–6 m, the distributor and the runner are directly submerged in a water chamber. Above this size, it is preferable to use a spiral housing to direct the water towards the turbine.

Spiral housing   The housing is a conduit which is spiral in shape and with a gradually decreasing section. It is designed to ensure that the energy of the water is divided evenly over the distributor.

At the junction between the housing and the distributor, there are often fixed vanes, or pre-guides, directing the streams of water to the inlet of the distributor guide vanes. The section through such a housing is shown in Fig. 34.

Distributor   The distributor is formed of a set of adjustable guide vanes arranged about a vertical axis, parallel to the rotation axis of the wheel, which passes through them at about the middle of their section. The guides are adjusted using small connecting rods controlled by a mechanism called the gating circle (Fig. 33).

The distributor has a dual role. Primarily it directs the streams of water towards the runner inlet in such a way as to avoid turbulence and to control the flow between full opening, giving maximum flow-rate and and complete cut-off (Fig. 33). The flow-rate corresponding to the maximum efficiency flow is about 0.8 times the maximum pressure for Francis and propeller turbines.

Draught-pipe   An exhaust conduit is installed at the outlet of a reaction turbine. It widens steadily (by 8–10°) towards the downstream side, and its purpose is to recover the residual kinetic energy ($C^2/2\mathbf{g}$) of the water and the head h_s between the exit from the wheel and the downstream water-level. This component is called the draught-pipe (see Fig. 34).

A simple calculation shows that the kinetic energy to be recovered increases with n_s. If no draught-pipe is included in a low-head installation, its efficiency will be seriously reduced.

Cavitation   While passing through the runner, the water pressure may fall below P, the saturated vapour pressure of the fluid, which for water is 22 mm Hg at ambient temperature. If this should happen, bubbles of water vapour and dissolved gas are formed which will strike the blades of the wheel and create vibration and parasitic turbulence. This can cause considerable damage to the machine and substantially reduce its efficiency. When the bubbles implode, the pressure can reach 300 bars and the frequency of bursting can be in the region of 1000 Hz, causing erosion and serious damage to the working parts of the machine. In order to avoid this phenomenon, precautions should be taken when selecting the head between the outlet of the susceptible area and the overall level. This is called the suction head (h_s).

A statistical study for various types of turbine with different specific velocities shows that the quantities log n_s and log σ, corresponding to the critical threshold, are sensitive indicators. Thus, the allowable region can be characterized by an equation of the type

$$\sigma \geq h_n n_s^\alpha$$

where α is called the Thomas coefficient.

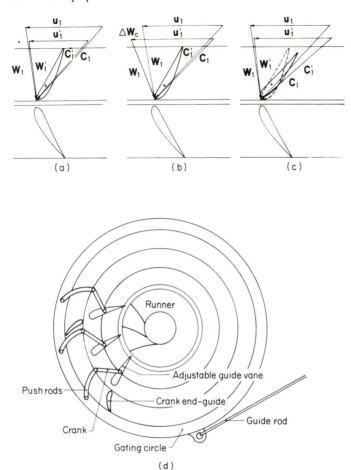

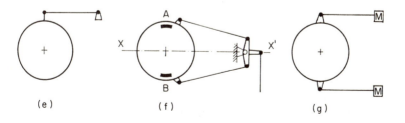

Fig. 33 Operation of the distributor: (a) fixed guide and speed control; (b) fixed guide and constant speed; (c) control of guide; (d) diagram of the principle of operation of guide vanes controlled by a gating circle; (e), (f), (g) different types of gating circle control

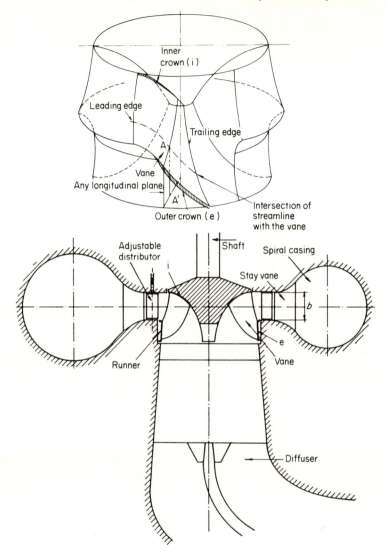

Fig. 34　Arrangement of vanes in a Francis turbine

This type of equation allows the foundation of the turbine unit to be sited relative to the downstream water-level and thus permits the civil engineering costs to be determined for the desired type of turbine. For Kaplan turbines, $\sigma = f(n_s)$ also varies with the operating conditions.

(b)　*The runners*

There is no fundamental difference between Francis and propeller runners, but there is a change in shape and size which takes place progressively from the Francis type to the propeller type as the specific velocity n_s increases. They can therefore be classified according to their method of operation as being of the propeller-centripetal Francis type or of the axial propeller type.

Francis turbines As can be seen in Fig. 34, a Francis runner has between eight and 15 blades, which are usually sited between an inner conical hub (i) and an outer cylindrical ring. These blades form twisted surfaces whose longitudinal section corresponds to an aerofoil and whose shape is determined from the specific velocity n_s.

This type of turbine is used mainly for heads of 10–100 m and for discharges up to 30 m³ s⁻¹. The rotation speeds are 250–1000 rev min⁻¹ and efficiencies are 80–90%, though they vary much more sharply about their optimum values as n_s increases.

Propeller and Kaplan turbines As n_s increases in propeller and Kaplan turbines, the flow of water through the runner becomes less radial until eventually it is completely axial ($n_s \geqslant 500$ rev min⁻¹). The relative velocities (**W**) become markedly higher than the absolute velocities (**C**) and it is preferable to replace the adjustable blades which channel the water flow with fixed ones. As the blades only slightly change the direction of the relative velocity, they become less numerous and the runner ring can be omitted. The Kaplan turbine is a propeller turbine with adjustable blades which permits turbulence-free operation under a variable head.

These machines are used for low heads ($\leqslant 10$ m) and large flows ranging from 5 to 100 m³ s⁻¹. Their efficiency is high (80–90%). A bulb unit for a propeller turbine on a low-power installation is shown in Fig. 35.

Axial thrust on impulse turbines In impulse turbines of either the propeller-centripetal or axial type, an axial thrust is produced by (a) the dynamic thrust of water masses diverted inside the moving runner, (b) the static thrust of the water on the various components of the runner, (c) the weight of the rotating components (if the machine is vertical), and finally (d) the axial thrusts in the conical reversing gear or spiral multiplier gear.

Calculation of these thrusts is important for establishing the containment forces.

4.6 CHOICE OF TURBINE

From the hydromechanical point of view, harnessing a head is a matter of determining the type and size of turbine to be used. As has been seen, hydraulic turbines are classified by family, and are characterized by specific velocity n_s (or Ω) (Fig. 36) and size, using some linear dimension such as the diameter (d_s) (or Λ). The optimum efficiency curves for a family of turbines vary with the specific velocity, as shown in Fig. 37. For discharge q, head h_n, and velocity n, the specific velocity n_s is given by equation (7).

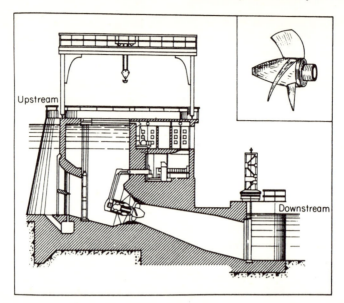

Fig. 35 A propeller turbine at Castet (SNCF, Pyrenees, France) having a power output of 810 kW with a head of 7 m. the system incorporates two bulb units in an open chamber designed to operate in a single chamber and each comprising

(a)   A propeller turbine of 810 kW capacity with adjustable blades and fixed distributor on a head of 7 m; speed 250 rev $\min^{-1}$; eight wheels, 1650 mm; eight bulbs, 1100 mm

(b)   A three-phase asynchronous alternator of 900 kVA, 500 V, 50 Hz, and 250 rev $\min^{-1}$, oil-cooled and placed upstream of the turbine in the water-stream

This was the first industrial upstream bulb installation. The bulb is inclined at 15° to the horizontal, the draught-pipe is straight, and the designs of the water circuits and the generator were very advanced. It is of unitary construction, and maintenance involves the complete replacement of the unit. The unit went into service in 1953

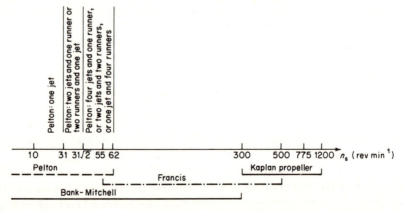

Fig. 36 Normal ranges of operation for different classifications of turbine as a function of specific velocity

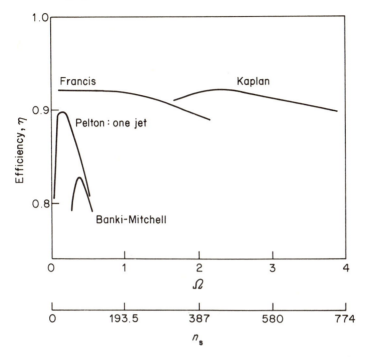

Fig. 37 Variation in the efficiency of different families of turbine with specific speed

The choice of machine results from a compromise between various factors such as the problems of cavitation, maintenance facilities, adaptation to large variations in the flow, etc., and ultimately the selection of the appropriate type of turbine for each head is founded on past experience.

Statistical analysis shows that the values of log h_n and log n_s for actual turbines of various types lie on families of parallel curves (Fig. 38).

Unfortunately this type of statistical data is not available for Banki–Mitchell turbines, and the curve given in Fig. 38, which has been established from the data which is available, is for guideline purposes only.

In the initial assessment of the project, this statistical data allows the specific velocity to be estimated for the turbine family corresponding to the head to be harnessed. Hence, for each harnessed discharge, it allows calculation of the rotational velocity, the size of the turbines, and, finally, for reaction turbines, the height (h_s) between the runner and the downstream water-level.

Finally, the efficiency chart for the family and the size of the installation allow the energy performance to be calculated. This permits an eventual analysis of the power, taking account of the discharges and their fluctuations, or of the choice of particular solutions such as using a multiple wheel. It should be noted that an increase in the size of the installation also increases

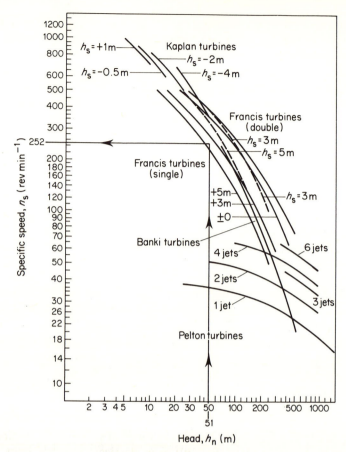

Fig. 38 Relationship between net head and specific speed for different families of turbine (from Vivier, 1964)

the amount of civil engineering work involved and thus the total capital outlay.

4.7 DEVELOPMENTS SPECIFICALLY RELATED TO MICRO HYDRO-POWER STATIONS

Over the last 20 years the exploitation of many low- or medium-power hydroelectric power stations has been neglected because of the relatively low cost of conventional thermal energy systems. As a consequence, they came to be regarded as uneconomic.

By contrast, the harnessing of high, or even very high, power sources has led to turbine companies developing the necessary technology. In such cases, the scale of the project merits the individual examination and design of each machine. This approach is not applicable to low-power installations where it is necessary to plan for minimum capital and maintenance costs and for the maximum possible period of operation.

Thus, the various turbine manufacturers have developed standardized units for small hydroelectric systems. Their designs are based on the following principles: (a) the optimum use of the latest research into turbo-machinery; (b) the supply of the electromechanical equipment in a compact ready-to-install and ready-to-operate form; (c) simple hydraulic design, using standard components to reduce costs and delivery times; (d) worldwide after-sales servicing.

This approach is applicable to powers between 100 and 2000 kW. Below 100 kW, capital costs become too high and the installation is therefore uneconomic. For powers over 2000 kW, individual design of the hydroelectric units can again be considered. Some small turbines, such as Hydrolec or Banki, which are particularly suitable for use in the developing countries without interconnected electricity grids, do allow the power to fall below 100 kW.

Similarly, the exploitable heads range from 1 m to about 400 m, with a rare extension up to 800 m. Following the earlier accounts, the standard types of turbine for this relatively wide range are (a) propeller (or Kaplan) turbines for low heads; (b) Francis turbines for medium heads; (c) Pelton turbines for high heads; (d) Banki–Mitchell turbines for a fairly wide range of heads; (e) some propeller turbines for very low powers ($\leqslant 100$ kW).

Large manufacturers have concentrated mainly on low heads so as to develop run-of-the-river equipment.

The solution most generally adopted for reasons of mechanical simplicity has been to develop adjustable blades in a range of standard diameters. These are mounted in siphon units (Fig. 39) or in classical axial flow units.

About 20 years ago, the Swiss company Escher-Wyss developed the manufacture of small right-angle bulb units. The alternator was mounted with its rotation axis perpendicular to the turbine axis. The French firms Neyrpic and Jeumont-Schneider took this idea further and developed a right-angle bulb-type unit with fixed blades and distributors. It is rigidly attached to an alternator which is fitted with its own exciter and control system (Fig. 40). In this way they have produced a compact, self-contained system which can be pre-assembled in the factory.

In the very low power range (up to 40 kW), the Leroy-Somer Company has developed a small compact unit called Hydrolec. It is a completely encased bulb unit with the casing acting as the draught-pipe. It can be mounted either at the end of a pipe or in a water chamber. This type of unit is shown in Fig. 41.

For coastal and estuarial installations, the tide-mill, in an improved form, can again be profitably revived for the supply of electricity in some characteristic sites, allowing for the application of small, standardized units which are commercially available.

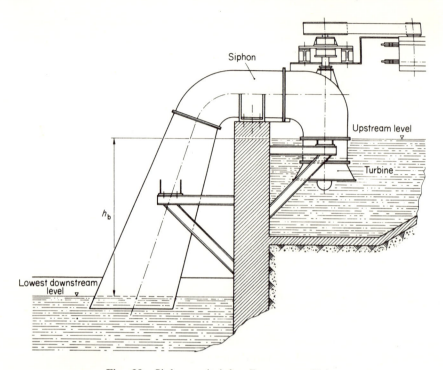

Fig. 39   Siphon unit (after Dumont et Cie)

Standard turbines can be applied most easily in installations with powers of up to 100 kW, with heads of between 1 and 400 m and for discharges ranging between 0.1 and 20 m³ s⁻¹. This is also the preferred range for micro hydro-power stations.

Figure 42 shows the power ranges of the various types of turbine in terms of the head and discharge.

4.8   SPEED REGULATION OF TURBO-ALTERNATOR UNITS

In the ideal case, power generators should satisfy the following two conditions: (a) the voltage supplied should be constant in amplitude and frequency, regardless of the supply conditions and any temporary variations to which it may be subjected (see Section 4.9.2); (b) the waveform of the supply voltage must, as far as possible, be sinusoidal, with negligible harmonics, particularly at radio frequencies.

In practice, for economic reasons, the criteria must be relaxed somewhat. The pattern of demand in an isolated network is generally quite varied, as it

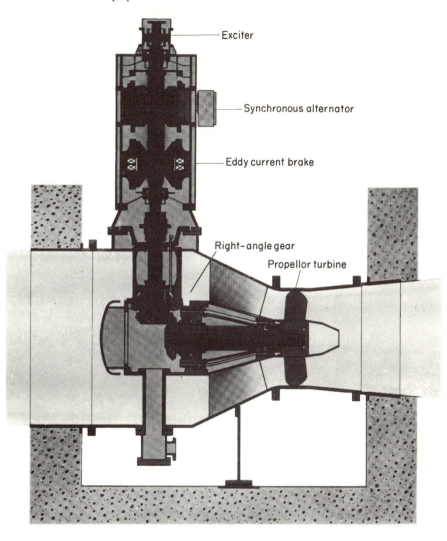

Fig. 40 Standard micro power bulb unit fitted with its own alternator and control system (from Neyrpic)

can include heating and lighting systems (which are affected only by the amplitude of the supply), mechanical loads (generally provided by asynchronous motors which are affected by the amplitude and frequency of the voltage, as well as creating demand for reactive power (see Section 4.9.3)), and telecommunications equipment which is very sensitive to interference from the electrical system.

Given these constraints, certain tolerances can be drawn up for the characteristics of the electrical energy supplied. Thus, the voltage should be

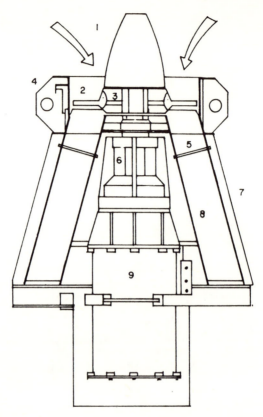

Fig. 41 Very low power Hydrolec unit: 1, nose-cone; 2, blade; 3, hub; 4, adjustable wicket gates; 5, stay vanes; 6, gearbox; 7, exterior cone; 8, interior cone; 9, asynchronous generator

stable to within ±5% whatever the demand; the power factor should always be greater than 0.8; the frequency should be stable to within ±2%; and harmonic distortion and radio interference should be kept to a minimum.

These constraints lead to the selection of generators with automatic control systems which react to demand variations. Speed control, which determines the frequency, is the most important and the most difficult to achieve.

4.8.1 Principles of speed regulation

If a turbo-alternator unit is subject to load variations on the network and is fitted with a control system and regulating equipment such as an inlet needle valve, passive resistances, eddy current braking, etc., then the variations in speed (and therefore in frequency) will decrease as the moment of inertia, I, of the rotating mass increases.

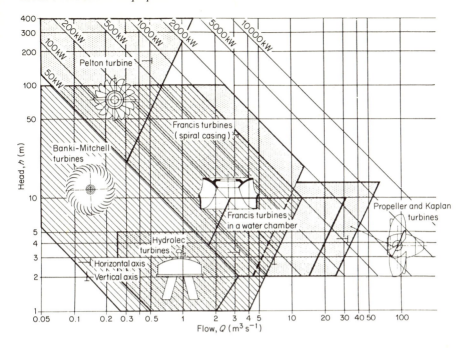

Fig. 42 Relationship between head and flow

This allows large power units (where I is very large) to have with long response times control systems such as the gating circle, the Pelton control needle, etc..

When light MHPS units are involved, these techniques become impracticable and electrical or electronic control systems with much shorter response times are required.

For medium-power units, a shaft-mounted flywheel is often used in order to increase I. (If there is a multiplying gear, it is preferable to mount the flywheel on the fastest rotating shaft.)

4.8.2 Mechanical regulators

These operate directly on the vanes, and are of two types. The first is a speed-sensitive controller such as a ball-governor, with a supplementary mechanism between the control and the regulation components to produce an initial stabilizing effect, followed by compensation as the speed returns to its original value (Fig. 43). The other type is a combined speed–acceleration governor which involves a tachometer to tell the controller the current speed $\Omega(t)$ to be reduced to Ω_0, and a damper to reduce oscillations.

As mentioned before, these mechanical governors are well suited to high-power systems, whose large inertia ensures good intrinsic stability, but

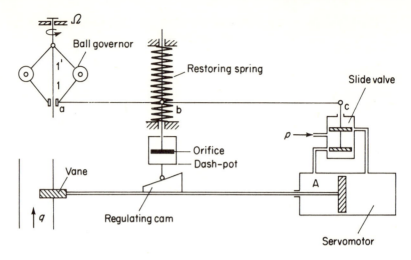

Fig. 43 Speed regulator with short-term control acting on the guide vanes

they are not suitable for low powers. Electrical governors are therefore preferred for MHPSs.

4.8.3 Electrical governors

These comprise a measurement sensor formed of an electronic frequency counter which controls the frequency of the transmitted voltage, and an amplifier to control an energy absorption system. This energy absorption is generally carried out by placing a battery of passive resistors in parallel with the load and subdivided so that they can vary almost continuously. Maximum power is always supplied by the alternator, and the surplus is consumed in the passive resistances. These resistances can be installed anywhere in the network, thus allowing optimization of the energy dissipated in the various operations, such as water hydrolysis, hydrogen production, heat production, etc.

The Jeumont-Schneider Company, in collaboration with Neyrpic, has developed a similar system, in which the excess energy is dissipated in an electric brake, mounted directly on the alternator shaft and in the same casing. The brake is designed as a homopolar engine with no winding on the stator, and the energy dissipation is through eddy currents in the solid stator which is cooled by water circulation. The eddy currents are generated by an exciter coil on an exciter stator, which is supplied directly by the power stage of the electronic regulator.

4.9 PRODUCTION, TRANSFORMATION AND TRANSMISSION OF ENERGY

4.9.1 Introduction

The mechanical energy available at the shaft of the turbine is not usually consumed on the site and so it is necessary to be able to transform it into a readily transportable and usable form. For reasons of safety and ease of transport, transformation and efficiency, an a.c. supply is normally used. With an alternating current supply, each parameter (current, voltage, or power) varies in a regular manner, changing its sign with a frequency f (in hertz) and a period T (in seconds). That is, these parameters have an angular frequency ω such that

$$\omega = 2\pi/T = 2\pi f.$$

At the other end of the power line, the consumption devices, which use the distributed energy, often require the frequency to be as stable as possible.

This implies that the distributed frequency must be held strictly constant. It must also have a stable amplitude and potential so that the current becomes the power-regulating factor. It will be recalled that a.c. power is expressed as $P = VI\cos\phi = S\cos\phi$, where P is the resistive power (in watts). Corresponding to the resistive power, the reactive power is $Q = S\sin\phi$, which is not actually a power but merely a convenient physical quantity which has been called power through analogy with P; it is expressed in volt-amps-reactive (VAr). S is the total power and is expressed in volt-amps; ϕ is the phase-lag between the current (I) and the voltage (V), and $\cos\phi$ is the power factor. The standard supply frequency is 50 Hz in Europe, and is 60 Hz in the USA.

Two general situations can arise which relate to these requirements for constant frequency, voltage and power factor. The first is when there is no grid connection or the available grid is too small. In this case, it is necessary to ensure that the installed unit is capable of providing its own regulation. This is the case with an isolated system. The second situation is when connection is envisaged to an outside electricity network which has a much higher power level than the planned installation (ratio greater than 10:1). This ensures the maintenance of a constant frequency and would cater for interconnection, for example, to a national grid to create a connected network.

4.9.2 Production of electricity for an isolated system

The power for an isolated system can be provided by synchronous machines, which provide alternating voltage at a given frequency, or by direct current generators which provide a continuous constant voltage.

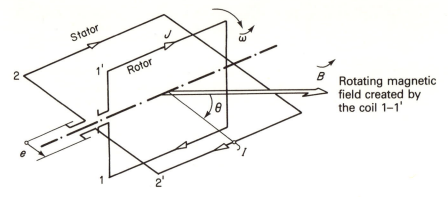

Fig. 44   Principle of the synchronous alternator

(a)   *Synchronous machines*

Principle   Assume a loop 1–1′ (Fig. 44) called the rotor, through which flows a continuous current of intensity J and which rotates at an angular velocity ω (the synchronous frequency) relative to a fixed frame called the stator. If this also carries an electric coil, each turn will cut the magnetic field created by the rotor, whose frequency is $2\pi/\omega = N$. An alternating current I of the same frequency will then be created in each turn of the stator.

A current is thus generated in the stator coil. By having three different stator coils and by interconnecting them correctly, three-phase current can be generated. The rotating element (the rotor) acts as the inductor producing the magnetic field. The fixed part (the stator) acts as the armature.

Two points are immediately obvious. The rotor speed must be stable as it controls the frequency, and a continuous current has to be supplied to the rotor for it to operate. The second point is simply solved by mounting another generator on the alternator shaft of the machine. This is called an exciter and its current is rectified and sent directly to the rotor windings. The first problem is much more difficult as the voltage and frequency of the current supplied, and thus the rotation speed of the unit, must be kept constant whatever the power demand of the network (see Section 4.8).

Rotation rate, frequency, poles   The alternator described above has an armature of magnetic material and is therefore bipolar. Also, the flux through each armature winding has the same value for each complete rotation of the inductor. Thus, the frequency of the flux and consequently the frequency (f) of the generated current is equal to the frequency of rotation (N).

Since, $f = 50$ Hz in Europe, the frequency of rotation of a bipolar alternator is 50 rev s^{-1}, or 3000 rev min^{-1}. Unfortunately, hydraulic turbines do not often rotate at this speed (see Section 4.5) so that one can either introduce a speed multiplier between the turbine and the alternator, as will be discussed in Section 4.10.1, or one can increase the number of poles on the inductor.

In an alternator with p pairs of poles, each phase of the armature comprises p coils separated by an angle of π/p. When the rotor has rotated once, the flux through each coil has experienced p of these periods and the frequency of the flux is thus p times larger than the rotation frequency, i.e. $f = pN$.

For $f = 50$ Hz, and thus $N = 50/p$ rev s^{-1}, one has

$$n = 3000/p \quad \text{rev min}^{-1}$$

In micro hydro-power stations, where the size of the installation, and therefore the amount of civil engineering involved, is to be kept to a minimum, the rotation speeds must be kept quite high, at least 500 rev min^{-1}, so as to keep the number (p) of pole pairs within reasonable limits.

The table below shows the rotation rates corresponding to 50 Hz for different numbers of pole pairs:

No. pairs of poles, p	1	2	3	4	5	6
n (rev min^{-1})	3000	1500	1000	750	600	500

For a frequency of 60 Hz (USA), these rates will be given by $n = 3600/p$, or 3600, 1800, 1200, 900, 720, and 600 rev min^{-1}. Typically, in a high-power, low-head unit, p can be as high as 64, and thus $n = 46.87$ rev min^{-1}.

Description   There are two types of inductor depending upon whether or not the poles are projecting (salient) or flush. In principle, machines with more than four poles always have projecting poles, two-pole machines always have flush poles, and four-pole machines can be of either type.

The various losses, such as Joule, eddy current, hysteresis, and mechanical losses, cause heating in the alternator. It must therefore be cooled to maintain a constant temperature and to protect the insulation under continuous operating conditions. When natural cooling is insufficient, the rotor can be fitted with a fan to provide forced air circulation.

Figure 45 shows a 10-pole synchronous alternator ($p = 5$) of Chinese manufacture. The projecting poles, the fan on the rotor, and the coils on the stator can all be clearly seen.

Losses and efficiency   There are two types of losses in a synchronous alternator. The primary losses are caused by electromagnetic (p_f) and mechanical (p_m) phenomena which occur when the machine is in operation.

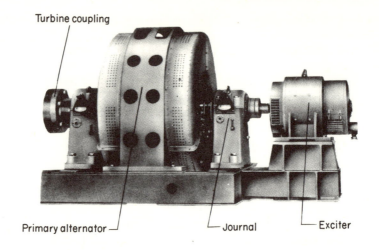

Turbine coupling

Primary alternator — Journal — Exciter

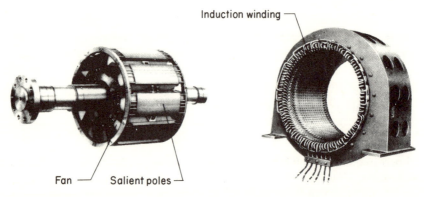

Induction winding

Fan — Salient poles

Fig. 45 A 10-pole synchronous alternator (from CMEC, Peking)

They are principally the Joule losses in the copper of the stator and rotor windings, iron losses in the stator, friction losses in the bearings, rings and brushes, and the losses due to the fan. The supplementary losses are caused by electromagnetic phenomena independent of the operation and fall into open-circuit and short-circuit categories. They are caused by the dispersion flux in the stator, harmonics in the magnetomotive forces and cogging of the rotor and stator.

The efficiency of a synchronous alternator can be calculated from

$$\eta_{al} = 1 - (\Sigma p)/(P + \Sigma p), \tag{9}$$

where P is the useful power and Σp is the sum of all losses.

For fan-cooled alternators with powers between 0.5 and 3000 kW operating at full load with cos $\phi = 0.8$, an efficiency of between 92 and 95% can be anticipated. For more powerful alternators, these efficiencies can reach 95–98%.

(b) *Direct current generators*

Although direct current cannot be widely used because of the difficulties that it presents for transformation and transmission, it is still used in some industries, such as electrometallurgy and electrochemistry, and for traction in cable-cars, elevators, railways, etc. It is therefore of interest to look briefly at d.c. equipment and its potential.

Description and operation Consider a rotor formed of a single turn coil 1–1′, as shown in Fig. 46, connected to a split-ring commutator. If the coil rotates in a constant and uniform magnetic field, it will produce a voltage at the terminals of the sliding brushes which is continuous in direction but which is not constant. This potential is shown as a function of the time in Fig. 46. If n turns are now arranged uniformly on the rotor and are connected to a rotation ring with n elements, a continuous and almost constant voltage will appear between the terminals of the two brushes. In order to create the magnetic field, a continuous winding is arranged on the stator through which flows part of the current produced. This is called the excitation current.

For generators, one of the three arrangements represented in Fig. 47 is most often used. They are fitted with a field rheostat which controls the magnetic field.

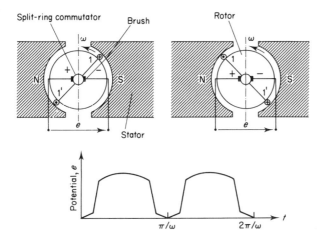

Fig. 46 Principle of the d.c. generator

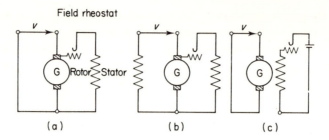

Fig. 47 Methods of excitation of d.c. generators: (a) shunt excitation; (b) compound
excitation; (c) separate excitation

Energy balance in d.c. generators; efficiency Consider a d.c. generator. To
generate electric power, it must rotate in the direction opposite to the
electromagnetic forces. It must therefore be provided with a mechanical
power P_m at its shaft to overcome the electromagnetic torque $N_e = P_e/\omega$,
where P_e is the electromagnetic power. Hence, the electric power $P = VI$
supplied at the terminals of the machine is equal to the electromagnetic power
$P_e = EI$ reduced by the Joule losses RI^2 in the armature.

The following illustration shows the general distribution of the powers in a
d.c. generator. From this, the overall efficiency can be defined using the same
equation as was used for alternators (equation (9)):

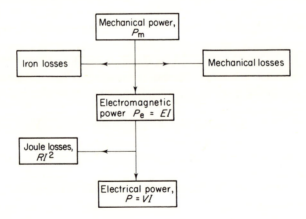

The sum of the losses is generally between 2 and 6% of the mechanical
power. The efficiency of a d.c. generator is therefore between 94 and 98%.

4.9.3 The production of electricity for a connected network

Interconnection should be considered if access is available to a high power
network of at least 10 times the power being installed. The outside grid will

impose its frequency and voltage on the smaller installation. If such connection is envisaged, an asynchronous machine should be installed.

(a) *Principle of the asynchronous machine*

Consider a magnetic field rotating at angular frequency ω, produced by an alternating current flowing through one winding on the stator. If the rotor is formed of a single short-circuited turn, then a current will flow through it in a direction such that it tends to oppose the motion that created it (Lenz's law).

The rotor therefore begins to rotate in the same direction as the induction current until it reaches a stable frequency ω' which is always less than ω, the synchronous frequency. Indeed, if $\omega = \omega'$, the flux through the winding will be constant and the induced current will be zero because the induced torque will balance out the loss torque.

As the turn seems to slip relative to the rotating field, the 'slip' is defined by

$$g = (\omega - \omega')/\omega.$$

Thus, an asynchronous motor rotates at a speed

$$\omega' = \omega (1 - g),$$

which is always less than the synchronous speed. This is the principle of the asynchronous motor.

From a manufacturing point of view, asynchronous machines are similar in many ways to synchronous machines, at least as far as the stator is concerned. In general, the rotor is a wound coil whose terminals can be connected to each other or to three rings on the shaft. A three-phase rheostat can be inserted in series with the rotor winding, allowing the torque to be increased on start-up while reducing the current by increasing the rotor resistance. Asynchronous machines should have an even number of poles on both the rotor and the stator, but the rotor can be wired for any number of phases.

Very often, the coil windings can be replaced by copper or brass bars, arranged in slots. The front sections of the coils only serve as connectors, and they are replaced by two identical, very low resistance rings, called shorting rings. This is called a squirrel cage rotor and is the type of structure normally used for generators.

(b) *Operation of the machine*

By applying a driving torque N_m to this machine, the synchronous speed can be exceeded, so that the new rotor speed becomes $\omega'' > \omega$. The torque of the asynchronous machine is then exerted in the direction opposite to the motion, and it becomes resistive. The machine now absorbs mechanical energy and transforms it to electricity, which is transmitted to the network to which it is connected. The machine operates as an asynchronous generator. The slip has become negative with $g = (\omega - \omega'')/\omega$. The frequency ω'' is called the hypersynchronous speed.

In both cases, the magnetic circuit must receive magnetization energy in order to produce the flux. This energy is always supplied to the machine from an external source and it is reactive because it corresponds to a current component which is orthogonal to the voltage. To avoid penalizing the network providing this magnetization energy (cos ϕ too low), a battery of capacitors is connected in parallel with the machine to provide at least some of the required reactive power.

(c) *Energy balance in an asynchronous machine*

The following illustration shows the general balance of powers in an asynchronous machine. N_s is the torque on the shaft, N_t is the transmitted torque, P_u is the useful electrical power, Ω' is the hypersynchronous speed ($\Omega' = \omega'/p$), Ω is the synchronous speed ($\Omega = \omega/p$), and $P_s = N_s\Omega'$ is the shaft power.

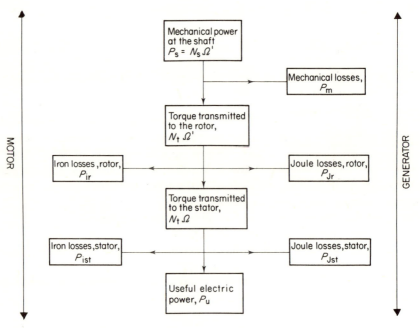

The energy balance for the machine is

$$P_s = P_{ist} + P_{Jst} + P_{Jr} + P_{ir} + P_m + P_u.$$

Hence,

$$P_s = \text{losses at the stator} + N_t\Omega - N_t'\Omega' + P_m + P_u.$$

When these losses are deducted, the stator transmits or absorbs the power $N_t\Omega$ depending on whether it operates as a generator or as a motor.

It can be seen that

$$N_t(\Omega - \Omega') = P_{Jr} + P_{ir} = N_t((\Omega - \Omega')/\Omega) = N_t g$$
$$= (P_s - P_{Jst} - P_{ist})g.$$

As the stator losses are very small in comparison with P_s, we can write $P_{Jr} + P_{ir} \simeq gP_s$. However, the iron losses in the rotor can be regarded as small since the radial induction wave turns slowly with the rotor ($g\omega$). Consequently, $P_{Jr} \simeq P_s g$. Thus, the Joule losses in the rotor are proportional to the slip.

Finally, the efficiency of a generator is expressed, using equation (9), by

$$\eta = 1 - (\Sigma p)/(P_u + \Sigma p).$$

This produces $P_u + \Sigma p = P_s$ and $\Sigma p \simeq P_{Jr}$, from which one has

$$\eta \simeq 1 - P_{Jr}/P_s \quad \text{where} \quad P_{Jr} \simeq gP_s.$$

Hence, $\eta \simeq 1 - g$.

Thus, the efficiency of an asynchronous machine improves as the slip decreases. An approximate idea of the efficiency of an asynchronous machine can be obtained simply by measuring its rotational speed.

An example of the calculations for an asynchronous machine and a particular application is given in Summaries 8 and 9.

Summary 8 Example of calculations for an asynchronous machine

---

Consider a unit with a rating plate showing four poles, 1475 rev min^{-1}, 50 Hz, $V = 380$ V, cos $\phi = 0.85$, $I = 140$ A.

It can immediately be determined that the synchronous speed for two pairs of poles is 1500 rev min^{-1}. Thus

$$g = (1500 - 1475)/1500 = 0.017 \Rightarrow \eta \simeq 1 - 0.017 = 0.983;$$

i.e.

$$\eta \simeq 98.3\%$$

The generator frequency is given by $1500 (1 + 0.017) = 1525$ rev min^{-1}.

One can also calculate the value of the capacitors to be connected in parallel to provide the reactive power required for the magnetization of the machine:

$$P = VI \sqrt{3} \cos \phi = 78\,500 \text{ W};$$

$Q = VI \sqrt{3} \sin \phi = 48\,700$ VAr.

The capacitance needed in a delta-connection is obtained by

$Q = 3V^2C\omega \Rightarrow C = Q/3V^2\omega; \qquad C = 357$ μF.

The capacitance needed in a star-connection is obtained by

$Q = 3U^2C\omega$

with $U = V/\sqrt{3}$ (single phase), and thus

$C = Q/U^2\omega = 1070$ μF

Summary 9   Example of using an asynchronous machine as a power regulator

---

Consider a turbine T on a head which fluctuates widely and which thus provides power varying from 120 to 200 kW according to the season. The turbine is to supply a machine-tool which requires 160 kW. To do this, a 40 kW asynchronous unit is connected to the turbine–machine-tool system. Because the unit is asynchronous and thus operates sometimes as a motor and sometimes as a generator, it is possible to supply a constant 160 kW to the machine-tool:

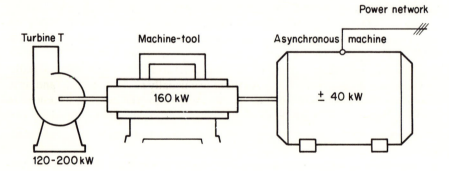

This example shows the benefit that can be derived from such equipment, provided, of course, that a suitable power network is available.

4.9.4   Transformation and transmission of the energy

(a)   *Transmission*

The energy produced is rarely consumed on the site. The power must therefore be transmitted to the consumer site while keeping line losses to a minimum.

The problem is thus as follows. If the characteristics (e.g. the absorbed power) of the supply equipment and switchgear are known, then it is necessary to determine the cross-section of the conductors which will allow a current I to be carried on a cable of length L with a maximum voltage drop ΔV in a network with power factor $\cos \phi$.

As the voltage V between two phases at the end of the line and the power P to be carried are known, the impedance of the line can be calculated using

$$Z_{max} = \frac{\text{allowable } \Delta V}{KIL},$$

where K is a constant equal to 2 in a single-phase system and $\sqrt{3}$ in a three-phase system, and

$$I = P/(V \cos \phi).$$

Z_{max} is the impedance per unit length of line. If, by way of simplification, the line is taken to be purely resistive, then $L_{max} \simeq \rho_c/s$, where ρ_c is the resistivity of the conductor and s is the cross-section of the cable. Thus, for given P, $\cos \phi$, and L, and with an allowable voltage loss ΔV, the cable cross-section is inversely proportional to the rated voltage. The amount of copper used for cable, and thus the cost, can therefore be reduced substantially by increasing the transmission voltage.

This calculation does not allow for any temporary surge, for example on motor start-up. For such cases, it is assumed that $\Delta V/V$ can have a maximum value of 10%, and that the starting current of the motors is three to five times the nominal run-current stated by the manufacturer. The calculation is made for these worst-case conditions. As there are usually several motors, and as it is unlikely that these will all start at the same time, the maximum power is generally taken to be that of the largest motor on start-up, together with the sum of the rated powers of the other motors. At start-up, it is generally assumed that $\cos \phi \simeq 0.3$–0.5.

When making simple estimations, it is generally assumed that $\cos \phi \simeq 1$ for filament lamps or heating elements, 0.8–0.9 for motors under normal operating conditions, 0.85 for compensated fluorescent lamps, and 0.3–0.6 for non-compensated fluorescent lamps.

In Europe, the normal inter-phase voltage is 380 V for low voltage transmission and 20 kV for medium voltage transmission. Low voltages cannot be used if the line is more than 1 km long, and in all other cases, especially when grid-connection is desired, the medium voltage levels must be used. These can be produced directly by the alternator from the outset.

Finally, the efficiency of a transmission line, just as for any other electrical appliance, can be defined by

$$\eta_{li} = 1 - \Sigma p_{li}/(P_u + \Sigma p_{li}),$$

where Σp_{li} is the sum of the line losses and P_u is the useful power at the end of the line. The efficiency can be expressed as a function of the allowable decrease in voltage by

$$\eta_{li} = 1 - \Delta V/V; \qquad \text{for } \Delta V/V = 10\%, \qquad \eta_{li} = 90\%.$$

(b)   *Transformation of the voltage*

As the cost of transmission is inversely proportional to the inter-phase voltage, it is often worthwhile being able to raise the voltage before transmission.

Alternating current simplifies this problem by permitting the use of transformers, which are simple stationary machines, and which are therefore robust and highly efficient.

A transformer basically consists of two magnetically coupled electrical coils with as large a mutual induction between them as possible. To achieve this, a magnetic core is used which is generally formed of plates, closed over on itself, and carrying the two coils.

If a sinusoidal voltage V_1 is applied to one of the coils (called the primary winding), a sinusoidal voltage is induced across the other winding (called the secondary winding). If a circuit is connected across this secondary winding, energy will be supplied to it.

The transformer thus absorbs energy through one of its windings and restores it through the other. For a perfect transformer with no losses, it can be shown that the transformer ratio is $m = n_2/n_1$, where n_2 is the number of turns in the secondary winding and n_1 is the number of turns in the primary winding, and if

$$m > 1, \quad \left\{ \begin{array}{l} V_2 > V_1, \\ I_2 < I_1, \end{array} \right. \qquad \begin{array}{l} \text{voltage-raising transformer,} \\ \text{current-reducing transformer.} \end{array}$$

$$m < 1, \quad \left\{ \begin{array}{l} V_2 < V_1, \\ I_2 > I_1, \end{array} \right. \qquad \begin{array}{l} \text{voltage-reducing transformer,} \\ \text{current-raising transformer.} \end{array}$$

By exchanging the primary and secondary windings, the same transformer can be used in both cases.

As the transformer has no moving parts, its efficiency is always excellent (between 90 and 99%). The only losses to be considered are the Joule losses in the two windings, which are variable and given by $P_j = R(I_2)^2$, and hysteresis and eddy current losses (P_v) in the plates of the magnetic core. These losses are constant for V_1.

The energy balance of the transformer can be shown schematically by the following diagram:

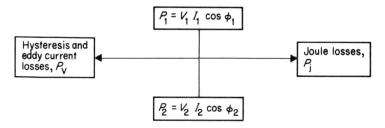

The efficiency is given by

$$\eta = 1 - \frac{P_v + R(I_2)^2}{V_2 I_2 \cos \phi_2 + P_v + R(I_2)^2},$$

or more simply

$$\eta = \frac{V_2 \cos \phi_2}{V_2 \cos \phi_2 + (P_v/I_2) + RI_2},$$

where V_2 is constant. Thus, if η is a maximum, $P_v/I_2 + RI_2$ must be a minimum; hence $P_v = R(I_2)^2$.

Thus, the following general relation can be derived. The efficiency of a transformer is a maximum when the constant losses are equal to the variable losses. This result implies that a component must be sized to suit the application if maximum efficiency is to be achieved.

4.10   OVERALL ENERGY BALANCE OF A HYDROELECTRIC POWER STATION

In order to draw up such a balance, the problem of combining a turbine and an alternator should be considered, although this cannot be done directly in most cases.

4.10.1   Turbine–alternator transmission gear

The need for a standard range of turbine runner diameters in MHPSs, designed for various heads, implies that the rotation speeds will vary greatly depending on the application. This can conflict with the selection of a suitable alternator as these are normally designed for fixed speed operation.

To solve this problem, it is often necessary to introduce a velocity multiplier between the turbine and the alternator. The most commonly used multipliers are as follows:

(a) Gear multipliers: these are not recommended for small powers because of their high cost and low gear ratios. For high powers, however, they provide the only possible solution given that the dimensions of the geared wheels increase with the power to be transmitted. In principle, their efficiency is excellent.

(b) Chain-wheel multipliers are limited by the linear speed of the chain, and they are therefore often unsuitable.

(c) Flat belts and pulleys have very low efficiencies because slippage between the belt and the pulley can be considerable.

(d) Vee-belts and pulleys are an excellent solution in most cases (at least for powers of less than 1000 kW), and permit a ratio of 4 : 1.

4.10.2   Overall energy balance

Examination of the various components of electromechanical equipment for a micro hydro-power station has resulted in the drawing up of an energy balance and an efficiency for each component. The efficiency for the whole installation will represent, therefore, the ratio between the net power (P_n) supplied to the turbine and the power (P_u) usable at the end of the line. The difference between these two values represents (a) pressure or hydraulic losses, related to the load losses in the turbine; (b) losses through flow leakage in the turbine; (c) electrical losses through the Joule effect in the conductors; (d) magnetic losses through hysteresis and eddy currents in the various magnetic circuits; (e) mechanical losses caused by the various rolling and thrust bearings in the rotating machines, turbulence in the turbine, the gear system, the energy absorbed by the alternator fan, and the energy consumed by the various regulating components such as the jack, the gating circle, the needle valve, etc.

The various losses between P_n and P_u are outlined in Summary 10.

By including the energy absorbed by the regulating components in the mechanical energy of the turbine, the following overall efficiency is obtained:

$$\eta = P_u/P_n = \eta_{tu}\eta_{gr}\eta_{al}\eta_{tr}\eta_{li},$$

where η_{tu} is the efficiency of the turbine, η_{gr} is the efficiency of the gear, η_{al} is the efficiency of the alternator, η_{tr} is the efficiency of the transformer, and η_{li} is the efficiency of the line.

Allowing for the various results obtained above, we can estimate $\eta_{tu} = 80\%$; $\eta_{gr} = 98\%$; $\eta_{al} = 97\%$; $\eta_{tr} = 98\%$; $\eta_{li} = 90\%$; and thus the overall efficiency of the installation is $\eta = 67\%$. Thus, an efficiency of about 0.7 can be anticipated for a well-designed and well-equipped installation.

Finally, Summary 11 gives two diagrams summarizing the general principles for the production of energy in both isolated and connected circuits.

Summary 10 Overall energy balance for a hydroelectric installation

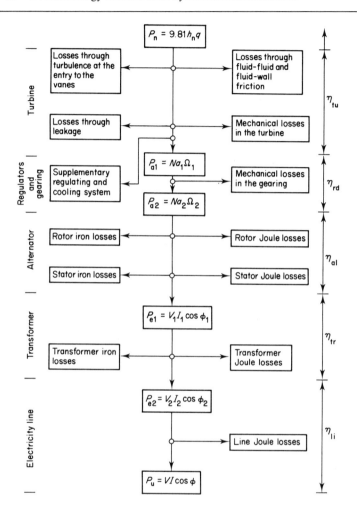

P_n is the net power of the turbine; P_e is the electrical power; P_s is the mechanical shaft power; P_u is the useful power at the end of the line.

4.11 EXAMPLE

The following example illustrates the use of the calculation techniques developed above.

Consider a watercourse with a hydrological pattern which has a wide seasonal variation, with an ensured minimum discharge of about $0.6 \text{ m}^3 \text{ s}^{-1}$ and a maximum discharge greater than $4 \text{ m}^3 \text{ s}^{-1}$.

The net head (h_n) has been taken as 51 m, and the problem is to determine the characteristics of the machine or machines which are best suited to large variations in the load.

Summary 11   General principles for the production of energy in an isolated
installation and one connected to a grid

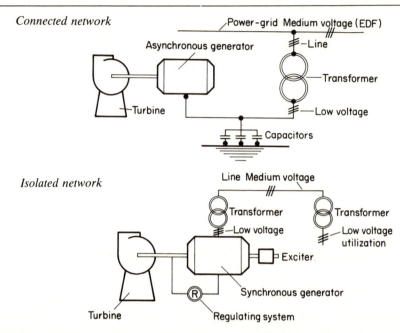

NB   For an installation in which the transmission is over a short distance, the low to medium
voltage transformer can be omitted. This occurs, for instance, with an isolated installation such as
a mill.

Reference to Fig. 38 shows that a net head of 51 m corresponds to the
application range of Francis turbines. Thus, taking a suction height (h_s) of
3 m, a Francis-type turbine with a specific velocity of $n_s = 252$ rev min^{-1}
would be appropriate.

From Fig. 29, the minimum discharge q_{min} corresponding to an efficiency
$\eta > 0.8$ is given by 0.5 q_{max}, or if $q_{min} = 0.6$ m^3 s^{-1}, $q_{max} = 1.2$ m^3 s^{-1} and as
$q_{optimum} = 0.8\ q_{max}$, $q_{optimum} = 0.96$ m^3 s^{-1}.

Thus, a wheel has been identified which can absorb a flow ranging from 0.6
to 1.2 m^3 s^{-1} with a minimum guaranteed efficiency of 0.8, providing an
optimum power of $P_n = 435$ kW. Equation (7) therefore allows the rotation
speed of the machine to be calculated from its specific velocity n_s and

$$n = n_s\ (h^{5/4}/p^{1/2});$$

i.e.

$$n = 1345 \text{ rev min}^{-1}.$$

This speed is very close to the synchronous speed for a three-phase
alternator (1500 rev min^{-1}), which can therefore be connected directly
without employing any gearing.

As this speed is fixed by the conditions needed for the operation of the alternator, it can be used as the reference when determining a second turbine for the larger flow range.

By further requiring that this second wheel can absorb twice the optimum (rated) flow of the first wheel, then $q_n = 1.92$ m^3 s^{-1}, where $P_n = 865$ kW (1020 hp) and its specific velocity is evaluated as

$$n_s = 312 \text{ rev min}^{-1}.$$

From Fig. 29, such a Francis turbine has a guaranteed efficiency $\eta > 0.8$ for a discharge ranging from $q_{min} = 0.69$, $q_n = 1.92$ m^3 s^{-1} and $q_{max} = q_n/0.8 = 2.40$ m^3 s^{-1}.

Finally, by operating the two wheels in parallel (a double Francis wheel), an optimum discharge of $q_n = 1.92 + 0.96 = 2.88$ m^3 s^{-1} or a power of $P_n = 1200$ kW can be absorbed for a specific velocity (n always equals 1345 rev min^{-1}) equal to $n_s = 374.6$ rev min^{-1}, with $\eta > 0.8$ over a flow range from $q_{min} = 1.92$ m^3 s^{-1} to $q_{max} = 3.60$ m^3 s^{-1}.

Thus, a machine has been designed which is formed of two Francis wheels mounted in parallel on the same shaft, rotating at a rate of 1345 rev min^{-1} and capable of absorbing a discharge ranging from 0.6 to 3.60 m^3 s^{-1} with a head height of 51 m.

The calculation of the sizes for each of the wheels should be carried out from the model turbine efficiency charts produced by the turbine manufacturers.

Consider the efficiency charts for Francis wheels which have the peak of their isospecific velocity curves at about 252 and 312 rev min^{-1}. Given n_s, η, and q/q_{max}, these diagrams yield values of N_{11} and P_{11} for the model turbine. From the relation $N_{11} = ND/H^{\frac{1}{2}}$, we derive $D = N_{11}NH^{\frac{1}{2}}$, the diameter of the actual wheel. The values of the powers and the efficiencies as a function of the degree of opening (q/q_{max}) can always be specified in this way from these efficiency charts.

The calculation should be completed by determining the required shaft diameters for the maximum power to be transmitted ($P_{max} = 0.8 \times 51 \times 9.81 \times 3.60 = 1440$ kW) to ensure that it does not overly obstruct the outlets of the runners in relation to their own dimensions.

BIBLIOGRAPHY AND FURTHER READING

ANONYMOUS (1979). Les microcentrales hydro-électriques dans le département du Puy-de-Dôme. Unpublished document, DDA, Département du Puy-de-Dôme, Clermont-Ferrand.

BACHMANN, J. (1980). Standardizing small turbines. *Water Power and Dam Construction*, **32** (7), 40–42.

BERNARD, J. and MAUCOR, S. (1980). *Microcentrales hydrauliques*, Editions Alternatives, Paris.

CAUVIN, A. and GUERRÉE, H. (1968). *Éléments d'hydraulique*, Eyrolles, Paris.

CORBEL, H. (1980). Aperçu sur la production autonome d'énergie hydro-électrique. Unpublished document, Centre technique Génie rural Eaux et Forêts, Antony.

GINOCCHIO, R. (1978). *L'Énergie hydraulique*, Eyrolles, Paris.

INDACOCHEA, E. (1979). Problematica del desarrollo de la tecnologia de micro-centrales hidroelectricas y su contribucion a la electrificacion rural. Report from the Instituto de Investigacion Tecnologia Industrial y de Normas Tecnicas, Lima, Peru.

INDACOCHEA, E. (1979). Hydroelectric micropower stations in Peru. Paper presented to the UNIDO seminar workshop Katmandu, Nepal, September 1979.

KOSTENKO, M. and PIOTROVSKI, L. (1969). *Electrical Machines*, Vols 1 and 2 [in Russian], Mir, Moscow.

LEFORT, P. (1969). *Les turbomachines*, PUF, Paris.

MOCKMORE, C. A. and MENYFIELD, F. (1949). The Banki water turbine. Report from the Engineering Experiment Station, Oregon State University.

NOYES, R. (ed.) (1980). *Small and Micro Hydroelectric Power Plants*, Noyes Data Corporation, Park Ridge, N.J.

OUZIAUX, R. and PERRIER, J. (1972). *Mécanique des fluides appliquée*, Vol. 1, Dunod, Paris.

SEDILLE, M. (1966–70). *Turbomachines hydrauliques et thermiques*, Masson, Paris.

SMITH, N. (1980). Histoire de la turbine à eau. *Pour la Science*, 1980 (29), 29–35.

TENOT, A. (1932). *Turbines hydrauliques et régulateurs automatiques*, four volumes, Eyrolles, Paris.

WALDEZ BAEZ, L. (1968). Estudio sobre microsistemas hidraulicos. Report from the Instituto de Investigaciones Electricas, Mexico City.

VARLET, H. (1964). *Turbines hydrauliques et groupes hydro-électriques*, Eyrolles, Paris.

VIVIER, L. (1964). *Turbines hydrauliques et leur régulation*, Albin Michel, Paris.

The reader is also referred to the technical and commercial literature of the following companies: Neyrpic, Jeumont-Schneider, Dumont et Cie, Leroy-Somer, Charmille, Ossberger, China National Machinery and Equipment Export Corporation, and Kossler (see Table 13).

CHAPTER 5

Project economics: capital investment and profitability

5.1 PROJECT ECONOMICS: CAPITAL INVESTMENT AND PROFITABILITY

In Europe during the 19th century much use was made of the power from small hydro-power stations. This allowed electrification to develop economically in outlying regions prior to the establishment of national electricity distribution grids. The widespread acceptance of this interconnection has unified the energy distribution system and has usually resulted, for reasons of ease of management and economies of scale, in the development of large electricity power stations managed by a single organization.

This concentration of plant was made at the expense of small installations for supplying local communities or private individuals. The fall in the cost of electricity with the development of oil-fired power stations had the effect of freeing consumers from the hazards of producing hydroelectricity themselves and from the costs of maintaining plant which was becoming less able to cope with the demands of the consumers.

The rise in energy costs over the past decade has stimulated a reassessment of the economic and social values of decentralized hydroelectricity production. Thus, a sound economic foundation can be provided for whole regions in fairly mountainous areas if small rural communities and small industries can take advantage of an energy-producing technology which does not require fuel and which involves easily maintained equipment with a long lifetime. Hydro-power has become an economic tool for the development of such disadvantaged regions.

For similar reasons, there has been an expansion in the installation of MHPSs in the developing countries where they meet the requirement for widely dispersed, low power electrical installations. A comparison with diesel installations today is generally in favour of the micro hydro-power station as this requires less maintenance and can be supervised by non-specialist personnel. The fuel is also free.

Nevertheless, there are several possible choices available, both for the machinery involved and for the type of use and the application. These choices determine the conditions controlling the profitability of the project.

5.2   CONDITIONS FOR THE USE OF MICRO HYDRO-POWER STATIONS

A hydraulic turbine can supply energy in four forms: heat, electrochemical energy mechanical energy and electricity, and the type of energy produced depends upon the power available. In the developed countries, hydraulic turbines with powers of less than 30–35 kW are suitable only for domestic or private use. With powers of between 30 and 150 kW, they can be constructed for domestic use, small-scale industrial or commercial use and the supply of current to the national grid. For powers greater than 150 kW, the energy has to be transformed into electricity because consumption at this level is rarely required at the turbine site. For powers of up to a few hundred kilowatts, on-site use is possible (e.g. for supplying ski-lifts run by small communities or for a water-purification plant, etc.), but above this level some or all of the electricity produced has to be diverted to the national grid.

For small communities, however, production for local consumption is of much greater importance than supplying the national grid, as a locally available supply for small industries and for the local population can create or sustain local employment activity, whereas electricity produced for sale to the national grid by a private producer is of insignificant economic value to the

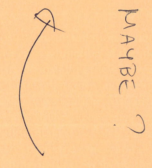

INSTALLATION

d by the discharge and the available power. The figures in Table 6 indicate ns. In general, for a given power, a than the equivalent high-head plant. l per installed kilowatt can be as much greatly improved, however, if the m which already exists. Even if large nent cost can be greatly reduced by of earlier watermills as the financial ydro-power station can be installed es.

relative costs. For example, the micro larger project such as a tourist amenity

The capital investment may have to be financed by loans, which can

Table 6 Breakdown of capital investment (in French francs at 1979/80 prices)

Country	Head, H (m)	Power, P (kW)	Price per installed kilowatt (French francs)		
			Electromechanical equipment and turbine	Civil engineering	Total
Low heads					
France	< 5	20	6000 (35–40%)	60–65%	16 000
USA/France	<10	50	4800 Rarely more than 30% of the total price for low-head systems	60–80%	12 000–16 000
USA/France	<10	200	3000		
France	3	>500	5000–6000		12 000
France	4	>500	2500		
France	6–7	>500	4000 (2000 each for electro-mechanical equipment and turbine (40%))	60% (20% for power station and 40% for intake works)	10 500
France	10	>500			
High and medium heads					
Peru	50–60	20	83 000 (52%)	48% (with full-pipe)	16 000
France	50–200	150–400	3000–4800 (50–60%)	40–50%	6000–8000
France	50–200	200–1000	2500–4200 (50–60%)	40–50%	5000–7000
France	50–200	≥1000	2000–3600 (50–60%)	40–50%	4000–6000

Notes: The proportions given above represent average values based on actual examples. The particular conditions for an individual micro hydro-power station can change these figures completely.

Sources: UNIDO seminar workshop at Katmandu, September 1979; Neyrpic, France; Camille Dumont, France; Leroy-Somer, France; BRGM, France.

become the largest financial burden as maintenance and running costs are low, specialist personnel are not required, and the station has a lifetime of approximately 30 years.

5.4   CRITERIA AFFECTING THE ECONOMIC VIABILITY OF A MICRO HYDROELECTRIC POWER STATION: INCOME FROM PRODUCTION

5.4.1   Profitability criteria

The economic criteria used by a large-scale power utility are obviously drawn up on a national scale. The national utilities are required to make an economic choice between the different types of power station, such as nuclear, fossil, hydro, etc., and to allow for the political consequences of such a choice. These consequences include such considerations as dependence on energy imports, the effect on the balance of payments, and the development of technology which is suitable for export.

For an independent producer or a local community, the economic criterion is the return on investment in terms of the evolution of the cost of purchasing electricity from the national grid (including the income from energy sales to the grid). Thus, the savings are determined in terms of both the savings which will be achieved by independent generation and the income which will be obtained by selling to the national grid. The viability of an independent generating project is thus related indirectly to the criteria used by the electric utilities.

The present tendency is towards reducing the price of base load power (off-peak) and increasing the price of energy at critical peak times during the winter. This would appear to be unfavourable for run-of-river power stations with no storage reservoirs as they cannot artificially modulate their production.

The value of the economic viability study for a run-of-river micro hydro-power station thus depends increasingly on the accuracy of the hydrological study used to determine the pattern of usable flows throughout the year. It can be interesting to verify the profitability of already installed MHPSs and to compare the actual production levels with the theoretical levels estimated when the project was being constructed. It is quite usual to find that independent producers run into difficulties in loan repayment after the initial period of enthusiasm.

Figure 48 shows a plot of the distribution of annual production against installed power for 100 micro hydro-power stations. For clarity, many of the points have been combined and only the average result for the group has been reported.

The micro hydro-power stations represented here are 'economic' in that they have been in operation for many years. However, the curve for a

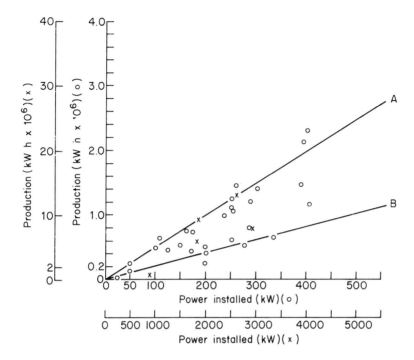

Fig. 48 Average production for existing viable installations. Lines A and B are the theoretical values after 5000 and 2000 hours of operation respectively. (Data applies to the Massif Central, Aquitaine, and the Pyrenees.)

theoretical operation of 5000 hours, which is often used as the viability criterion, forms the upper boundary of the cluster of points. Thus, it must be admitted that the potential of micro projects is generally overestimated. This occurs through insufficient knowledge of the discharges and through basing the calculations on a series of hydrological data which is too short or which is non-representative of the catchment area. On this point, it should be remembered that the cost of taking a daily flow measurement for a year is less than 0.7% of the total investment cost of the project and that these measurements, by extension of the data, allow an accurate evaluation of the hydraulic potential of the site to be obtained (see Chapter 2).

5.4.2 Income from a micro hydro-power station

Over the 10-year period from 1971 to 1980, electricity charges in France have typically risen by over 15% per annum. The figures for EDF are shown in Table 7. Nuclear power will affect this trend, but it is not clear to what extent.

Table 7 Evolution of price per kilowatt-hour for sales to EDF

Year	Winter		Summer	
	Price (FF/kWh)	Annual rate of increase (%)	Price (FF/kWh)	Annual rate of increase (%)
1971	0.0613		0.0316	
1973	0.0737	+ 10	0.0361	+ 7
1974	0.0906	+ 23	0.0440	+ 22
1975	0.1012	+ 12	0.0539	+ 22
1976	0.1212	+ 20	0.0690	+ 28
1978	0.1469	+ 11	0.0879	+ 14
1979	0.1768	+ 20	0.1021	+ 16
1980	0.2035	+ 15	0.1114	+ 9
1981	0.2605	+ 28	0.1201	+ 7.8
1982	0.3027	+ 16	0.1234	+ 2.7
1983	0.3282	+ 8.4	0.1187	− 3.8

The correct choice of the economic criteria to use for a private micro hydro-power station depends largely upon how the energy crisis will evolve over the next 20–30 years, the lifetime of the installation. This, of course, is speculation on what will happen in the future, but some guidelines can be drawn up to clarify the subject a little. Depending upon the application, the income from a micro hydro-power station is formed in part by the savings in energy purchases and in part from selling power to the national grid.

In France, the tariffs for purchases made by EDF are based on one fixed rate per kilowatt for guaranteed powers and five rates per kilowatt-hour for power supplied at different times of the year. These five tariffs are peak time rate, winter rate, off-peak winter rate, summer rate and off-peak summer rate.

The winter period is from 1 October to 31 March, and supply for complete hours during the critical period between November and February is paid for at the peak time rate.

There are two tariffs for purchases, the 'integrated', and the 'simplified' or single tariff. The integrated tariff distinguishes between supplies of power guaranteed by the producer for at least 5 years on pain of a reduction in the fixed rate and partially guaranteed supplies which do not involve a fixed rate.

Until 1977, the prices per kilowatt-hour were calculated from those of the guaranteed supply, adjusted by factors defined in the scale of charges:

	Winter	Summer
Programme supply	1	0.8
Diagram supply	0.8	0.7
Intermittent supply	0.5	0.5

Programme supply corresponds to production which is agreed one week in advance, diagram supply is that which is agreed on the previous day, and the balance is the intermittent power.

However, to encourage this type of electricity production, legislation in 1977 raised the prices of all of the partially guaranteed energy to the level of programme supply, and there are now only two factors (1 in winter and 0.8 in summer).

The simplified tariff was introduced by EDF for powers of less than 500 kW. The tariff scales can have two, four or five tiers. By way of example, the simplified tariff established in 1980 was as follows (in French centimes, exclusive of VAT):

	Five-tier	Four-tier	Two-tier
Critical hours, winter	35.80	27.81	22.79
Peak hours, winter	26.21		
Off-peak hours, winter	15.27	15.27	
Peak hours, summer	13.07	13.07	11.41
Off-peak hours, summer	8.91	8.91	

The most common tariff is the two-tier one, but the owner of a dammed system may apply for the four- or five-tier tariff. Furthermore, if the simplified tariff does not contain a fixed premium, the prices benefit from increases in quality of up to 12% in winter and 20% in summer.

Thus, in 1980, the income from the sale of electricity to the French national grid could be estimated from the installed power, the number of hours of production and their distribution:

Installed power (kW)	Income through sales to EDF (French francs)		
50	From	15 000	to	43 000
100	From	30 000	to	85 000
200	From	60 000	to	170 000
500	From	150 000	to	425 000
1000	From	300 000	to	850 000
3000	From	900 000	to	2 500 000
5000	From	1 500 000	to	4 250 000

5.4.3 Annual running costs

The principal items in the annual running costs are as follows:

(1) Building attendance and running maintenance costs such as cleaning the grill (0.4–1.2% of the initial investment, I)
(2) Sundry charges such as payments for water rights for the head-race or the full-pipe (0.8% of I)
(3) Operating licence (at least 0.4% of I)
(4) Insurance of the electromechanical equipment and the turbine (1% of I)
(5) Provision for repairs (1.2% of I).

During the first year these expenses are about 4% of the total initial investment, and they increase by about 10% per annum, excluding interest repayments.

5.4.4 Economic cash flow

Cash flow is given by revenue (inclusive of all taxes) minus operating costs and VAT throughout the lifetime of the installation. It allows the internal rate of return to be evaluated for the project.

The internal rate of return is the discount rate at which the sum of the discounted values of the cash flow is equal to the investment (Summary 12). It is generally assumed for a micro hydroelectric power station that it should be greater than 15%.

Summary 12   Some definitions in calculating the project economics

---

Discount rate

This allows the present value of money to be compared with its value in n years time. This concept allows a relationship to be established between present and future worth: 1 money unit in n years time corresponds to $1/(1 + a)^n$ today, where a is the discount rate.

Discounted profit

An investment (I) over n years produces a record of expenditure (E) and receipts (R):

$$\text{Discounted profit,} \qquad \bar{P} = -1 + \sum_{t=1}^{n} \left(\frac{R_t - E_t}{(1 + a)^n} \right) ;$$

$$\text{Discounted expenditure,} \qquad E = +1 + \sum_{t=1}^{n} \left(\frac{E_t}{(1 + a)^n} \right) .$$

Depreciation period

Number of years after which the sum of the discounted receipts becomes greater than the sum of the discounted expenditure.

Pay-back period

Number of years after which the sum of the non-discounted receipts becomes greater than the sum of the non-discounted expenditure.

Rate of return

This is the value of a which produces a zero discounted profit, P:

$$\bar{P} = -1 + \sum_{t=1}^{n} \left(\frac{R_t - E_t}{(1 + a)^n} \right) = 0.$$

Note: In financial matters, the discount rate is replaced by the interest rate.

5.4.5 Financial profitability

The annual pre-tax profit is obtained by deducting interest, depreciation and loan repayments from the cash flow. Depreciation is calculated from the total amounts exclusive of VAT at the rate of 5% for the civil engineering, 7% for equipment, and 10% for other costs.

5.4.6 Financial cash flow

Cash flow is the after-tax profit after allowing for depreciation. If it is greater than the annual capital repayment instalments and other expenses, then a profit is produced.

The profit thus depends upon three factors – the variability of the turbined flows and of their distribution in time, the variability in the tariffs, and the evolution of the price per kilowatt-hour.

For each project, it is also necessary to examine the permitted ranges of variation in these three, and sometimes more, factors which will allow the financial balance of the operation to be preserved even during years when production is low.

Indeed, a series of low production years can cause a delay in the ability to repay loans and can lead to financial difficulties. The hydrological study should allow this risk to be reduced, while not overlooking its importance.

5.4.7 Examples of trading accounts

An example of a set of trading accounts is given below:

	$P = 20$ kW	$P = 500$ kW	$P = 2200$ kW
Capital Investment	300 000 FF	1 620 000 FF	5 500 000 FF
Receipts	17 000 FF	330 000 FF	900 000 FF
Expenses:			
Provision for repairs	1 500 FF		
Insurance	1 900 FF	93 000 FF	137 000 FF
Sundries	900 FF		
Maintenance	10 000 FF		
Interest	8 000 FF	46 000 FF	
Profit before depreciation	−5 300 FF	191 000 FF	763 000 FF
Depreciation	18 000 FF	105 000 FF	135 000 FF
Profit		86 000 FF	628 000 FF
Profit per kilowatt		172 FF	285 FF

Thus, in practice, a private micro hydro-power station of 10–20 kW is profitable only if it is constructed on a 'do-it-yourself' basis on a site which does not require any extra civil engineering work.

This type of installation can be financially viable, however, in some of the developing countries, where the costs and operating difficulties of a diesel installation would be prohibitive (see Section 5.6).

5.5 METHODS OF FINANCING AN INSTALLATION

Various financial incentives to encourage energy saving have been instituted and in France, for example, energy policy was entrusted in 1982 to a special energy agency (l'Agence Française pour la Maîtrise de l'Énergie). As part of its remit, this department is charged with encouraging small power stations, particularly to benefit local authorities. Subsidies are granted for diagnostic and feasibility studies of potential MHPS sites, and, under certain conditions, grants-in-aid for especially large undertakings can be as much as 30% of the overall construction costs.

Local authorities can also effect loans from a more general fund.

5.6 MICRO HYDROELECTRIC POWER STATIONS IN THE DEVELOPING COUNTRIES

The priorities are very different in the developing countries, where present programmes for the electrification of rural areas without access to a national network depend upon diesel generators.

Diesel electric units and large hydroelectricity production capacity can coexist. Thus, countries such as Ghana with the Akosombo power station, Nigeria with the Kanji power station, Pakistan with the Tarbela power station, Zaire, Gabon, etc., have sufficient hydraulic potential to be able to contemplate partial or total supply of their electricity demand through high-power hydroelectric stations. Nevertheless, a low population density scattered over a wide area, which is sometimes further complicated by climatic and geomorphological difficulties as in Zaire and Gabon, cannot justify the extension of an electricity distribution grid into the rural areas as the cost is insupportable. Thus, the supply of electricity to rural areas relies upon low-power stations which are dependable and require the minimum of maintenance.

The following figures quoted at the Lomé Conference, May 1979, and at the Katmandu Conference, September 1979, give examples of the plant capacities required:

(a) Supply for a small village (500 inhabitants) with one school – 50 kW
(b) Supply for a medium-sized village with one school and a medical centre – 100 kW
(c) Supply for one tea or cotton factory during the day and one village during the night – 200 kW.

These installed capacities were regarded as sufficient and, indeed, as suitable, for encouraging a stable population by raising the standard of living and by allowing economic expansion in the village which would then become a centre for the surrounding area.

Table 8 Comparative costs in 1979 for 1 kWh of electricity produced by low-power electricity power stations (5–30 kW) using various energy sources in the Third World countries

Interest rates (%)	Cost per kilowatt- hour (in French francs)			
	Connection to an existing grid at 25 km distance	Diesel electricity generator	Solar panels photovoltaic cells	Hydroelectricity
10	1.59	1.29*	1.37	0.43
5	1.05	1.23*	0.95	0.30
2	0.78	1.21*	0.73	0.24

*Oil price assumed to be 1.13 FF per litre.
Source: Conference at Lomé, Togo, 1979.

Table 8 shows that since 1979 the micro hydroelectric power station has been a less expensive means of producing electricity than using solar or diesel units, or connecting to an existing grid further than about 25 km away.

The advantage of micro hydroelectric power stations in the developing countries arises from the low rates of interest charged by the international financing agencies, which thus encourage the installation of equipment with a high capital cost but with low running costs.

The importance of diesel units has decreased greatly in recent years. The advantage of low initial investment has been nullified by the rise in running costs (mainly, the cost of oil) and the other disadvantages of diesel units which cannot be overlooked. These include frequent breakdowns, necessitating trained maintenance staff, and difficulties in oil deliveries due to seasonal variations in the state of the road network or long distances from the distribution centres.

The micro hydroelectric power station is therefore now seen to be a very competitive method of electricity production in countries with widespread hydraulic resources. Good examples of these developments include China, Indonesia and Peru, which countries are currently expanding this effort by making the maximum possible use of local resources for labour and materials to create economic activity in the villages concerned.

BIBLIOGRAPHY AND FURTHER READING

ANONYMOUS (1979). Application de la mini-production d'énergie hydro-électrique au pays en développement et ensembles énergetiques pour les régions rurales du tiers-mondes. Paper presented at the Symposium interrégional des Nations Unies sur le Processus de Développement et les Options technologiques dans les Pays en Voie de Développement, 21–26 May 1979, Lomé, Togo.

ANONYMOUS (1968). Appréciation de la rentabilité économique des investissements: note de méthode provisoire. Working document, Ministère de l'Économie et des Finances, Paris.

BAUDOUIN, G. (1979). Production hydroélectrique décentralisée. *Annales des Mines*, 1979 (April), 57–60.

BIENVENU, C. (1976). Les énergies nouvelles. *Revue générale africaine*, 1976 (10), 25–29.

BONAITI, J. P. (1979). Les impactes socio-économiques de la création de petits aménagements hydrauliques. Report from the Institut économique et juridique de l'Énergie, Grenoble.

COTILLON, J. (1979). Micro power: an old idea for a new problem. *Water Power and Dam Construction*, 1979 (January), 42–48.

DUMON, R. and CHRYSOSTOME, G. (1980). *Les pompes à chaleur*, Masson, Paris.

GLADWELL, J. S. and WARNICK, C. C. (1979). *Low-head Hydro: An Examination of an Alternative Energy Source*, Idaho Water Resources Research Institute, Moscow, Idaho.

INDACOCHEA, E. M. (1979). Contributions to the UNIDO seminar-workshop on the exchange of experiences and technology transfer on mini hydroelectric generation units, 10–14 September 1979, Katmandu.

LAMBERT, Y. (1979). Le soleil pour le feu et l'eau. *Afrique–agriculture*, 1 September 1979, 20–27.

LEENHARDT, O. (1977). Quand la psychologie l'emporte sur l'économie et les finances. *Reforme*, 27 August 1977, 79.

SKOULM, O. (1980). Les microcentrales hydro-électriques et le milieu naturel. *Eau et rivières de Bretagne et basse Normandie*, 1980 (34), 18–21.

THUAULT, M. (1980). Les microcentrales hydrauliques. *Génie rural*, 1980 (June), 7–12.

UNITED NATIONS (1980). Report of the symposium on the prospects of hydroelectric schemes under the new energy situation and on the related problems, Athens, 5–8 November 1979. *Report no. ECE/EP/36*, United Nations, Geneva.

UNITED NATIONS INDUSTRIAL DEVELOPMENT ORGANIZATION (1980). Seminar workshop on the exchange on mini hydroelectric generation units, 10–14 September 1979, Katmandu. Report Extr. UNIDO, issue paper, pp. 13–18; *see also* Draft report on the UNIDO/ESCAP/RCTT joint meeting, pp. 22–32.

CHAPTER 6

Legal aspects and French legislation

Over the course of time, legislation has evolved concerning energy production. In France, extensive legislation was introduced in 1919 imposing conditions for the first time on the use of hydro-power. Electricity supply was nationalized in 1946 and, with very few exceptions, it specified the rights and duties of a new public body, Électricité de France. Measures to protect the environment were passed in 1976, introducing a requirement for evaluating the environmental impact of every installation. Recent legislation was introduced to encourage the production of electricity by means of MHPSs by permitting their installation in capacities of up to 4500 kW subject to a simple permit from the local authority and, for the first time, giving local authorities the right to derive financial benefit from this resource.

6.1 EARLY REGULATIONS ON THE USE OF WATERCOURSES AS AN ENERGY SOURCE

In feudal times, and then under the French monarchy, the use of water either for the production of mechanical power to operate watermills, sawmills or small-scale industry, or for irrigation and the provision of a public water supply, was the right of the nobility or the subject of a royal warrant. These property rights, and thus the rights to the use of the water, could be transferred to third parties.

Once these rights had been acquired they could be handed down from generation to generation going through the acts for the abolition of privileges passed during the French Revolution. Water intakes and any heads related to them were at that time regarded as 'titular' water rights or as having an existence in law.

In 1898 legislation was introduced specifying the distinction between 'navigable and floatable' rivers of relevance to the public domain, and others. That is, distinction was made between 'domanial' water, which is running water flowing on private property where at least half of the river-bed belongs to the proprietor, and 'non-domanial' water, where it does not form part of their property. In this latter case, there is a right of use to the water within the framework of the current legislation, but not of disposal or appropriation. It should be noted that a 'mixed watercourse' was also defined in 1964 where over half of the river-bed belongs to the riverains but for which the rights of use belong to the State.

Installations constructed on domanial watercourses prior to 1566 (*Édit de Moulins*, 1566) have been recognized as having inalienable titular rights. The same rights applied to non-domanial watercourses for all hydro-power installations established prior to the night of 4 August 1789, or which had been sold as state property during the Revolution.

It should also be noted that the legislation of 1919 upheld the titular rights of installations provided that the water intakes had not been modified and provided that no increase in the volume of the water flow was involved. The permanent right of titular holders of such property was confirmed, even though the installations relevant to the 'existence in law' only provided for a right which was limited in time.

Today, owners of such 'privileges' must provide proof of the ownership rights acquired through one of the following legislative means:

(a) An act of royal authority prior to the 1566 legislation containing the right to use the water in perpetuity
(b) A deed of sale of state property, also mentioning the right to use the water in perpetuity
(c) Proof of the existence by an uncontested fact of the extraction of water prior to the abolition of feudal privileges (4 August 1789).

These titular rights, granting the owners a positive administrative advantage, are at present under review in the light of the legislation of 1898. They can be changed by a simple administrative decision if it is proven that they form a hindrance to the flow of water or that they create an intolerable environmental impact, or that they have been forfeited through non-use, (e.g. where no trace of a previous installation can be detected).

Until the beginning of the 20th century, the utilization of a domanial river for the production of power was authorized, always subject to revocation, by a prefectorial judgment. On non-domanial watercourses, the water rights were recognized as being the property of the riverains.

Authorizations to extract water granted between the revolution and 1919 are valid until 1994, due to the stipulation in the 1919 legislation that they be upheld for 75 years. In 1994, those producers of hydro-power involved must apply for a new authorization which will be granted within the terms of the legislation then in force.

6.2   LEGISLATION RELATING TO THE PRODUCTION OF
ELECTRICITY

6.2.1.   Legislation of 16 October 1919

This legislation resulted from the anarchic development of the use of
hydro-power and the abuses on the part of the riverains or developers of this
new power source. The new regulation wished to safeguard the interests of
the public as the power derived from a watercourse is very much a national
asset. It therefore placed constraints upon the riverains so that they could
only use it within certain limits as authorized in the Civil Code.

The use of this new economic asset, hydro-power, was thus subjected to
some simple controls. It could be used under a licence granted by an order of
the Council of State for powers greater than 500 kW, or under a local
authority permit for powers up to 500 kW. In 1980 the threshold was raised to
4500 kW.

The legislation of 1919 dealt with the maximum gross power, i.e. no
allowance was made for pressure losses or the efficiencies of the electromechanical equipment.

6.2.2   Legislation of 8 April 1946

This law nationalized the electricity supply. The management of electricity
distribution was entrusted to a national public body of an industrial and
commercial nature to be called 'Électricité de France' (EDF). However, it did
allow the private sector the possibility of independent electricity production,
and thus companies whose average annual production for 1942 and 1943 was
less than 12 000 000 kWh were excluded from nationalization. An amendment
in 1949 permitted private individuals to install power stations with powers of
up to 8000 KVA (about 7000 kW). Furthermore, from 1955, EDF was
obliged to purchase electricity supplied by independent producers and
conditions were then specified for the purchase and sale of electricity together
with tariffs, which were to be adjusted from time to time.

6.2.3   Environmental protection

Legislation covering this was passed in 1976. Every installation should
henceforth be considered in terms of the impact that it would have on the
environment, and proposals for countermeasures should accompany every
application. The statement or the more detailed study of environmental
impact, depending upon the amount of power produced (see Chapter 7),
should cover four points:

(a) Analysis of the initial state of the site and its environment

(b) Analysis of the effects that harnessing the watercourse will have on the environment in terms of its natural beauty, landscape, flora and fauna, ecosystems, noise levels, etc., as well as the socioeconomic effects of the project

(c) Reasons for undertaking the project

(d) Measures envisaged by the contracting authority to suppress, reduce and if possible compensate for the harmful effects of the project on the environment, with an estimation of the total cost of these measures.

Summary 13   Statutory definitions of power

The methods for calculating power involve information and data supplied by the Hydraulic Consultative Committee in 1920 and 1926, following the promulgation of the law of 1919.

Maximum gross power

$$P = 9.81 \; QH,$$

where P is the maximum gross power in kilowatts, Q is the maximum flow at the intake in cubic metres per second and H is the maximum head in metres.

Maximum available power

$$P = 8Q \; (H - p),$$

where the pressure losses are p. The 9.81 is reduced to 8 to allow for the efficiency of the equipment.

Normal gross power

$$P = 9.81 \; qH',$$

where q is the average usable flow and H' is the head (i.e. the difference in height between the average water-level in the reservoir and the level where the water is returned to the flow).

The average flow used is not the average flow of the stream because it has to allow for flooding and the reserved flow downstream of the intake. The usable flows can be determined by studying the curve for classified flows in Chapter 2.

The average usable flow q over a large number of years, therefore, is

$$q = \frac{V \times 10^6}{31.5 \times 10^6} \quad \mathrm{m^3 \; s^{-1}},$$

where V is the annual usable volume in $10^6 \; \mathrm{m^3}$.

Normal available power

$$P \; = 8q(H' - p'),$$

where p' is the loss in pressure. The 9.81 is reduced to 8 to allow for the efficiency of the equipment

---

6.2.4   Legislation of 15 July 1980

In changing the law of 1919, this legislation raised the threshold for local authority permits from 500 to 4500 kW, and allowed local authorities to install power stations of up to 8000 kVA and to derive benefit from such electricity had to production. Until then, this possibility was not directly available to local authorities, as they could not be involved in profit-making activities. The production and sale of electricity had to arise as a by-product of another activity such as a water supply, irrigation, a leisure facility, etc. It now rests with the local authority to show how micro hydro-power stations can fit into a general economic development programme and how the financial resources released through the sale of the electricity to the national grid can be used to implement this programme.

6.3   THE TAKING OF EVIDENCE IN APPLICATIONS TO HARNESS WATERCOURSES FOR THE PRODUCTION OF ELECTRICITY

These procedures relate mainly to the above three laws of 1919, 1976 and 1980.

Authorization can be given in one of the following ways:

(a) A licence (*concession*), for installations with a maximum gross power greater than 4500 kW but less than 8000 kVA (about 7000 kW). Above 8000 kVA, the EDF monopoly on production, transmission, and distribution is exercised without restriction.

(b) A local authority permit (*autorisation préfectorale*) for MHPSs with powers below 4500 kW.

The procedure for obtaining a licence, as defined in 1960, is a long and complicated process, taking approximately three years. The processing of applications is handled by the Ministry for Industry and its regional boards. Several bodies are consulted at a local and central level, and the petitioner must prepare about 50 copies of the application. Inquiries are held in each of the communes which will be affected. The installation can be declared a public utility, which permits compulsory purchase orders to be issued and possibly the creation of obligations and constraints. When granted, the concession is for a maximum duration of 75 years. The installation for which the licence is granted becomes state property and is developed according to the regulations in the tender specifications approved in the licence order.

Summary 14 Licence application documentation

The licence application documents should include the following:

(a) A plan of the site on a scale of 1 : 50 000
(b) An outline plan of the proposed premises and construction works. The plan will show the perimeter of the concession area (inside which the obligations imposed in the modified law of 1919 will apply)
(c) A section along the watercourse showing the site works
(d) A description of any salient features of the works, financial studies, etc.
(e) A map of any land to be flooded, indicating the various types of cultivation and the total area of each
(f) The agreements which can be concluded between the petitioner and the local and national authorities, either from the point of view of finance or supplies of water or guaranteed heads, as well as agreements already made as to compensation in cash or in kind
(g) A table of the indemnities for unexercised rights that are proposed by the petitioner in favour of interested riverains
(h) Proposals for the distribution between interested communes of the rental from the head and its equipment
(i) A draft of the tender specifications, conforming to the model laid down in the 1919 legislation
(j) A statement of the typical gross and net powers available
(k) A list specifying the technical and financial capabilities of the petitioner, and proving that he fulfils the required nationality conditions
(l) The study of the environmental impact

The procedure for obtaining a local authority permit is the responsibility of the body charged with policing the water. It is quicker, usually taking several months, and only requires an appraisal by the water authority and a public inquiry. However, this permit is not permanent, and can be cancelled without prior warning or compensation should the MHPS become an impediment to any operation announced by a public utility.

In every case, the harnessing of the watercourse using an MHPS, whether by licence or permit, must be reconcilable with any other harnessing activity of EDF, whether under construction or planned for in the future.

Evidence in applications for a permit

As the power threshold has been raised from 500 to 4500 kW, the majority of MHPSs will now be covered by local authority permits. Not all of the governing regulations based on the law of 1980 have as yet been published, but the procedure, which is simpler than the licensing procedure, must be followed for new installations.

The application is directed to the local government administrator, the prefect, for the *département* in which the installation is to be sited. Depending upon the type of legal control governing the watercourse, the appraisal is entrusted to an appropriate authority such as the Ministries of Agriculture, Supplies or Transport. The application includes, in addition to the identity of the applicant, a certain amount of technical information relating in particular to the measures which will be defined by the local authority for regulating the water.

The investigating department checks that the application is admissible and can ask the applicant for further information, which must be provided within one month or else the application will be cancelled.

The application is passed for opinion to the various departments responsible, for example, for fisheries, architecture, the environment, and to the inter-regional industrial directorate which acts as a watchdog over EDF.

Summary 15   The permit

The permit comprises the following:

(a) The name of the watercourse in the *départements* and communes in which the installation will be sited, from the extremity of the back-wash to the outlet of the tail-race
(b) Plans for siting the installation, on a scale of 1 : 20 000 or 1 : 25 000:
   (i) Plan and elevation of the intake whose sides should, if possible, have depth indicators attached
   (ii) Longitudinal cross-section covering 200 m upstream and 100 m downstream and transverse cross-section upstream and downstream of the installation
(c) The principal market of the proposed electricity production, whether it is for consumption by the producer or for sale to EDF
(d) Salient features of the MHPS installation and electromechanical equipment, giving an outline description of the maximum gross power and the maximum available power, and the characteristics of the tail-race with the maximum flow to be carried
(e) Statement of existing hydro-power stations immediately upstream or downstream of the proposed installation
(f) A statement or study of the environmental impact, depending upon whether the power of the installation will be greater or less than 500 kW
(g) An outline estimate of the capital expenditure
(h) A note specifying the technical and financial qualifications of the applicant
(i) Expected lifetime of the installation
(j) Requested duration of the permit

Finally, the applicant should prove that he has French nationality or is a citizen under the jurisdiction of the European Community and that he has free use of the private land upon which the works are to be carried out.

The prefect deals in particular with those bodies with whom he is in direct contact – the committee for sites of natural beauty, chambers of commerce, agriculture, industry, etc.

A draft permit is prepared and presented to the prefect, who will then consult the elected General Council and the local population, in this case by means of a public inquiry, posters, and the opening of a register in the prefecture office and in town halls. For consultation purposes, the body entrusted with the policing of the water will draw up the final proposals.

The application for a permit can be rejected and the reasons for the refusal notified to the applicant. If the permit is granted, a copy of the order is sent to the ministry responsible for the control of waterways.

Agents of the department responsible for policing the water and for electricity supplies, as well as officials and agents authorized to report any infraction of water regulations, have access to the site at all times. They check that the conditions imposed upon the applicant are being satisfied. On completion of the work, the department confirms that the installation conforms to the regulations stated in the permit.

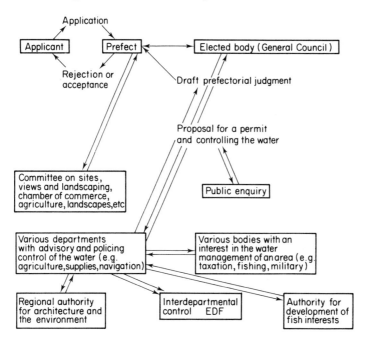

Fig. 49   Procedure for obtaining a permit

These procedures can be summarized as follows:

The national utility has a monopoly on the distribution of electricity, except for those plants and companies which were not nationalized. However, they do not have a monopoly on production at levels below 8000 kVA. Thus, in France, there is a legal right for the private sector to produce electricity, either for local consumption or for sale to the national grid.

EDF must buy the electricity produced by such independent producers and enter into a contract for at least the duration of the economic lifetime of the installation. It should also be noted that the producer can, for a certain charge, transmit the power he produces to other sites owned by him and resell his surplus.

Local authorities, who until recently only collected the business licence fees paid by the independent producers in their communes, now have access to production for use in their administrative offices and for sale to the national grid.

MHPSs can be authorized by two administrative procedures: a licence for installations with a gross output greater than 4500 kW but less than 8000 kVA (about 7000 kW), and a permit for outputs less than 4500 kW.

Applications are examined by the regional authorities and permits are granted by the prefect.

The conditions for the purchase of electricity produced by independent producers are fixed by law and the tariffs are changed from time to time. Modifications to the various laws were published in 1981 and can be found in the *Journal Officiel*, No. 94, 19 April 1981.

BIBLIOGRAPHY AND FURTHER READING

ARNAUD, P. and PONSOT, G. (1980). Contexte administratif de la production auton-
ome d'électricité d'origine hydraulique. Paper presented at ALPEXPO, 4e Salon international Aménagement en Montagne, April 1980.
GAZZANIGA, J.-L. and Ourliac, J.-P. (1979). *Le droit de l'eau*, Litec, Paris.
SOCIÉTÉ PROVENÇALE ÉQUIPEMENT (1978). Étude des possibilités de création de microcentrales hydroélectriques au bénéfice des collectivités de montagne. Internal report, Société Provençale Équipement, February 1978.
WEISENHORN, P. (1980). Rapport sur le projet de loi modifié par le décret relatif aux économies d'énergie et à l'utilisation de la chaleur. Assemblée Nationale, Session 1979–80, no. 1719, Annexe au PV, 14 May 1980.

CHAPTER 7

Environmental impact

The environment and damage to it are perceived in different ways by different groups of people. The environmental impact of a hydro-power station is not felt in the same way by the individuals who derive benefit from it as it is by those to whom it represents a nuisance.

The impacts of MHPSs can be assessed under the following headings:

(a) Awareness of the sensitivity of the natural environment
(b) Prediction and evaluation of the effects of the installation on the environment
(c) Formulation of methods for counteracting any harmful effects.

The statement of environmental impact (for $P < 500$ kW) or the study of impact (for $P > 500$ kW) included with the application for a permit will include details of

(a) The impact on the pattern of the water system, the aquatic ecosystem, the landscape, the existing noise levels in the area and the socioeconomic impact mentioned in Chapter 5
(b) The reasons for choosing the MHPS site
(c) The measures proposed to suppress or reduce, and if possible to compensate for, the harmful consequences of the planned installation (together with an estimate of the costs of repairing the damage).

7.1 VISUAL INTEGRATION

The various components of an MHPS installation can be integrated with the landscape to differing degrees. Too often, however, the aesthetic aspect of these installations is neglected by the installer, who is insensitive to the eyesore he has created. A solution which is acceptable to all the parties involved can usually be reached with a little imagination and at a comparatively low cost.

145

An examination of the morphology of the site and its vegetation (forest, heath, etc.) will lead to an understanding of the various elements of the landscape into which the power station is to be inserted.

The various uses to which the land is put and the amount of human interference which has already taken place can be evaluated by considering various features of the site, such as the nature and type of habitat, the relationship between the habitat and the landscape and the extent to which the land is already being used for urbanization, industrialization, agriculture, leisure activities, etc.

The evaluation of the impact of an MHPS will take into account any landscaping schemes currently underway or planned for the various regional areas by the agencies responsible for planning, natural parks, protection of the environment, etc., and will also consider the appearance of the site from different positions to establish its visibility from roads, pedestrian footpaths, inhabited areas, etc.

It is often difficult to assess to what extent a particular environment is valued by its inhabitants or temporary users, but it is advisable to identify any attractive features such as rivers and lakes, woodland, the natural colours of the area or any historical features.

Countermeasures

The determination of the impact of the various elements of an MHPS installation upon the natural surroundings should allow measures to be formulated for their insertion into the landscape in such a way that a new landscape is created which is of equivalent natural beauty to that which preceded the installation.

The dam and the system for discharging flood water, the fish-pass and the inlet of the channel or full-pipe with its grill and grill-cleaner generally integrate well. A dam of earth or stones is more suitable for a rural site than a concrete structure. The designer should be conscious of all the natural or man-made features which combine to form the distinctive character of the countryside concerned, including the type of vegetation, the geological formations, the morphology and the coloration.

The head-race for the power station, the settling-pond and the forebay should not be difficult to integrate as the edges of these structures will quickly be softened with vegetation. As these particular elements are built on a hillside, they should blend in easily with the surroundings provided that there is sufficient plant cover.

In contrast, however, it is very difficult to overcome the aesthetic problems posed by the full-pipe. Builders usually limit themselves to running the pipe along the valley without taking any precautions and not even bothering to rearrange the boulders and rocks which have been displaced. When only the reserved flow is passing, the pipe appears as an ugly gash. It is therefore important to consider the path of the pipe in advance and to make the best possible use of those features in the surroundings which could provide good camouflage. It can also be painted a suitable colour for those stretches where total concealment by rocks, soil or vegetation is not possible or is too difficult.

The building housing the electromechanical equipment should be imaginatively designed. The entry of the full-pipe, the building itself, the tail-race and the electricity lines which run out from the building should be subjected to a design study to blend them into the surroundings in the most effective way. The granting of any building permits in future should be dependent upon an architectural presentation which has taken account of the recommendations laid down in the section of the notice or study of impact concerning the countermeasures to be adopted.

The electricity lines, even if they can be buried on leaving the power station, must eventually surface for reasons of cost, and they cannot then be easily integrated into the natural landscape. It is thus necessary to give thought to the selection of their path, and to the type and shape of the supporting poles.

The tail-race returning the water to the river also has to be integrated into the landscape in terms of both its shape and the type and colour of materials used to build it.

Those who will benefit from the electricity produced, whether private individuals or local authorities, should be aware of the responsibilities they incur for protecting the natural landscape when they undertake any MHPS project. It falls to them to plan and institute the vital measures to enable the whole installation to be satisfactorily incorporated into its surroundings.

7.2 NOISE IMPACT

The noise produced by an MHPS sited near dwellings can inconvenience those who live near the river. The noise is caused by various components of the electromechanical equipment, such as the turbine, the gearbox, the alternator and the transformer itself if there is one.

The flow of water in the turbine creates wide frequency-band noise which is related to the efficiency of the machine. On this point, it should be remembered that the efficiency of the turbine is dependent upon the turbulence at the entry to the blades, which forms one noise factor. In addition, there are various narrow-band components which are at frequencies proportional to the rotation rate. For example, the periodic passing of the wheel or propeller blades in front of the nozzle or the diffuser blades all produce components of ancilliary noise. To avoid resonance phenomena, it is also essential that the number of guides should be different from the number of paddles or blades on the runner. Other similar types of noise are caused by unequal pressures at different parts of the blades (or the vanes), causing speed fluctuations at the outlet of the wheel.

A harsh noise in the 1000–2000 Hz band can arise in the gearbox if there is any free play due to poor manufacture or excessive wear and tear. Indeed, the gearbox is often the main source of noise if it is badly designed or installed (e.g. through poor anchorage).

The alternator can also create noise. This can be mechanical in origin (as with air flow at the fan or noise in the bearings), or it can be electromechanical (such as that caused by vibrations in the magnetically saturated laminations of the stator).

The transformer, even though it is a static machine, can also be noisy because of the low frequency (50 Hz) noise arising in the laminations forming the magnetic circuit. For a transformer supplying a medium voltage line (50 kV), the noise levels vary with power as shown below:

P_n (kVA)	Noise level (dBA)	
5 000	59	← Upper limit of
10 000	63	power for MHPSs
20 000	67	
30 000	70	

The calculation of the overall noise level should include that of the flowing water, especially at the tail-race. This can be as high as 60–70 dBA for a channel of 1m × 1.5 m, equivalent to the noise of a waterfall on a small mountain stream.

Noise levels are usually set by regulation for various applications and are measured using noise meters according to standardized testing methods.

There are no regulations specifically covering micro hydro-power stations, except that, like any industrial installation, they must comply with any general recommendations on noise levels. Typically, these specify that a noise is deemed to be a nuisance during the day if it is 5 dBA above the background level (i.e. the noise level existing before the addition of the extra noise). During the night, the additional burden is typically reduced to 3 dBA. In practice, the tolerable noise level for a micro hydroelectric power station is 50 dBA at 10 m (Table 9).

Some steps for reducing the noise level can be undertaken at the time of installation. The electromechanical components, the chief source of noise, can be placed in a housing, which will give at least partial sound-proofing. Thus, lightweight sound insulation can be achieved economically and conveniently by using partitions of 700 g m^{-2} sheeting, coated in fireproofed PVC, with Perspex windows. Another very effective solution is to bury as much of the unit as possible. However, this is costly because of the terracing that would be required.

Table 9 Examples of noise levels

Noise level (dBA)	Effect	Examples
0	Threshold of audibility	
20	Very quiet	Rustling leaves, caves, undisturbed snowfields
40	Quiet	Residential area
50	Moderate	An office (the tolerable noiselevel for MHPSs at 10 m)
60	Acceptable but difficult for work involving thought	Conversation
80	Noisy	Radio at full power, railway station, street with heavy traffic
85	Noisy	MHPSs at 1 m
90	Threshold for injury to ears if suffered for 8 h per day	Mechanical workshop
100	Unpleasant	Underground railway station, motorcycle without a silencer at 2 m
120	Deafening	Pneumatic drill at 2 m
140	Unbearably painful	Jet aeroplane at take-off

7.3 BIOLOGICAL IMPACT

7.3.1 The water and the river

The water flowing in rivers is provided by atmospheric precipitation (see Fig. 3 in Chapter 2). It will appear immediately as runoff when it is raining and it will also appear more slowly after infiltration into the soil and subsoil, water flow in the aquifers and eventual return to the river through springs or more diffuse seepage.

In particular, most of the flow in a watercourse under low-flow conditions is provided by the groundwater and so, in summer, or in January and February in mountainous regions, when there is little or no rain, there is only groundwater flowing in the bed of the watercourse.

The water in the river provides aquatic life with all of its requirements: the oxygen which is vital for its survival, suitable temperatures for the various species to develop, suitable resting places, nourishment, physical support to allow movement, a cleaning agent permitting the removal of plant and animal waste, etc.

7.3.2 Disturbance of the water conditions

Whether it is calm or running, the water in a river provides the support for all animal and plant life. If variations occur in the pattern of the river or if the water quality is changed, aquatic life will be disturbed to an extent which is more or less surmountable depending upon the species involved.

Every artificial action, and particularly the installation of micro hydroelectric power stations, is liable to affect the flow of the water and to induce quality changes either by accelerating the flow upstream or by reducing the flow downstream. As a prerequisite, therefore, a good physical knowledge of the basin concerned and an astute evaluation of the repercussions of harnessing the river are required.

Upstream of the installation, there is an effect related, however slightly, to the water-level created by the intake structure. According to the nature of the dam, the water-table will undergo variations in level, which, in the event of flooding, can require drainage to be installed (Fig. 50).

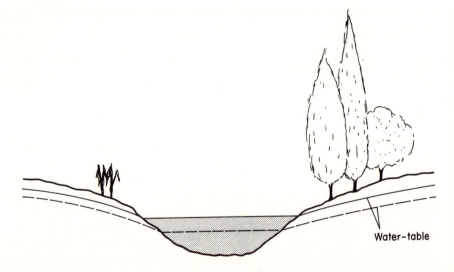

Fig. 50 Relationship between the water-level and the water-table

A further disturbance is caused by any increase in the amount of water evaporated from the water surface (related to its area, depth, insolation, wind, etc.), which can be as high as 6000–10 000 $m^3ha^{-1}yr^{-1}$.

These impacts are evaluated from the estimated volume and the area of the reservoir and from the maximum and minimum water levels before and after the construction, as these are related to the fluctuations in the water-table.

Downstream of the intake but above the tail-race outlet, the water flow is derived from (a) contributions from the groundwater in the aquifers in this part of the catchment basin; (b) water which has not passed through the turbine, particularly flood-water; (c) the reserved flow which was specified in the permit for the MHPS.

Over this section of the watercourse, the possible repercussions on the water-tables of any reduction in the flow must be evaluated from the hydrogeological data.

The section of the watercourse below the tail-race outlet can be affected when the water which has passed through the turbine is not all returned to the river. This occurs with MHPSs installed on a drinking water supply or on an irrigation system, or when water is diverted to another catchment area.

Thus, we again meet the situation of a section of the river with a reduced flow, with the ultimate consequence of a lowering of the water-tables with which it is connected. This situation requires knowledge of the extent of the relationship between the water-table and the river.

Typical operation involves the use of dams and sluice-gates which can temporarily modify the natural flow by a factor of three so that the frequency and duration of the gate openings and of the comparisons between the natural and modified flows is very important for evaluating the environmental impact.

7.3.3 River life

Under natural conditions, with no disturbance, healthy aquatic life is dependent upon (a) the water – its speed and volume flow rate, chemical quality, temperature and dissolved gases (oxygen in particular), and (b) the morphology of the river and its catchment basin – the gradient and width of the bed and the lithological nature of the bottom.

Even when there is a disturbance, plants and animal life may be able to adapt. For example, many species can gradually build up a resistance to chemical pollution through the process of mithridatism, and many more species would seem to have survived the large number of watermills which have existed on French rivers in the past, bearing in mind that France had 50 000 mills in operation at the end of the 19th century.

Plant and animal life are closely related. The vegetation provides the ichthyological fauna with some of its food, and often a place for rest and reproduction. Thus, aquatic macrophytes offer protection by permitting certain species to spawn and by allowing the biomass of aquatic invertebrates to feed and grow.

Various life forms (biocenoses), such as plants and vertebrate and invertebrate animals, can become established to an extent that depends upon the
natural environment (the biotope), its dimensions and the resources it can
provide to support life.

Fish and other life is thus distributed over the region according to
biochemical and biophysical criteria which allow the best conditions for their
fundamental physiological requirements – food, rest and reproduction. A
knowledge of this distribution allows the species to be identified which would
be disturbed by a particular installation.

Five distinct zones for the preferred development of fauna can be identified
by considering a watercourse from a point near its source, where the flow of
water in the torrent is relatively cold and well oxygenated, through stretches
where the river flows slowly with warmer water which has less oxygen, to its
estuary where the fresh water mixes with salt water (Table 10).

In practice, some species are ubiquitous and the passage from one biotope
to another rarely occurs with a sharp cut-off, but usually occurs over a
transition area, or ecotone, which is extended to a greater or lesser degree.

Certain migratory species travel long distances to reproduce. Some (the
catadromous fish) travel from upstream to downstream often as far as the sea,
and some (the anadromous fish) travel from the sea towards the upper
watercourse.

Thus, when an eel reaches maturity, it migrates from the river to the
Sargasso Sea to reproduce. The river lamprey travels up rivers, as do shad
and salmon. The Atlantic species of the Loire–Allier shad, originating in the
upper reaches of the system, will travel as far as Greenland before returning
to the river of its birthplace to reproduce.

Unlike these long-distance migratory fish, other species such as trout and
zander, and, to a lesser extent, pike and gudgeon, only migrate within one
river to lay eggs, feed and find shelter.

Trout perform a triple migration entirely within the river in which they
were hatched. When fully mature, they travel upstream to the spawning
grounds between mid-October and February or March. They then travel
downstream to rejoin their biotope, to be followed after hatching by the
young fish descending to the richer feeding areas. The reproduction seasons
correspond to low-flow periods (see Table 11), and, for fish in watercourses at
low and medium altitudes, this period, from March to June, generally
coincides with the electricity utility summer tariffs, the lowest tariff of the
year.

In contrast, for fish in high-altitude watercourses, and in particular for the
salmonidae, the reproduction season (from October to March) corresponds
to a period with a high electricity tariff and low water-levels in snow-fed
rivers.

Food requirements vary from one species to another and according to
whether the fish are adults or fry. Ranging from the solely vegetarian rudd to
the carnivorous pike, which even eat their own fry, the species include

Trout zone (mountain torrents, 5–10 °C)	Grayling zone (fast flowing rivers, 8–14 °C)	Barbel zone (medium-speed rivers, 12–18 °C)	Bream zone (slow rivers, 16–20 °C)	Flounder zone (estuaries)

Species distribution ranges across zones:

- Salmon
- Brown trout, rainbow trout, miller's thumb, minnow, brook trout
- Flounder
- Bleak
- Stone loach
- Burbot
- Rudd, Catfish
- Bream, Bouvière carp
- 10-spined stickleback
- Carp, zander, spined loach, Crucian carp
- Sun Perch
- Common grayling
- Huchon
- Chondrostome
- Soiffe (chondrostome), dace
- Tench
- Ruffe
- Grey mullet
- Chubb
- Freshwater bleak
- Shad
- Barbel, gudgeon
- Stickleback
- Roach, perch
- Pike
- Eel

Trout zone	Grayling zone	Barbel zone	Bream zone	Flounder zone
Well oxygenated running water; poor vegetation; strong-swimming fish with hydrodynamic shapes, e.g. trout, or with flattened bodies living near the bottom, e.g. chubb, phryganea larvae (e.g. caddis worms). mayfly and beetle larvae; food supply mainly of insects and larvae, molluscs, mussels, and plants	Well oxygenated water; beginning of sedimentation; root-establishment of plants	Oxygenated water; abundant flora on sandy and muddy bottoms; fish with high bodies, e.g. barbel, chubb, perch, etc.: limit of temperature for salmonidae, 18 °C	Moderately oxygenated water: submerged or floating aquatic plants allowing spawning, providing shelter and food for fry: cyprinoesocidae. e.g. cyprinidae. white carnivorous fish (perch, pike, zander, black bass, etc.): fish with high, flattened bodies, e.g. bream and tench	Intrusion of salt water

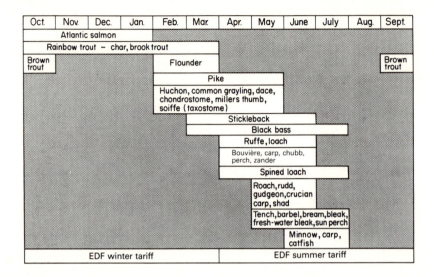

opportunist fish which, like trout, feed on benthic invertebrates and aquatic or land insects.

The fish-breeding productivity of rivers is very variable. Salmonidae rivers (first category rivers) produce insufficient trout to meet the requirements of anglers, whereas in cyprino-esocidae rivers (second category) with chub, roach, rudd, carp, bream, tench, bleak, etc., and carnivorous fish such as pike, perch, zander, black bass, etc., there are more fish than amateur and professional fishermen require (see Table 12).

First category watercourses as defined in France comprise the mountain torrents, plateau and hill rivers in Brittany, the rivers in the Champagne and Picardy plains, and total as much as 225 000 km. The productivity in terms of kilometre of bank per annum ranges from 1.5 (for torrents) to 50 kg km^{-1} yr^{-1}.

Second category watercourses, which include all rivers other than the salmonidae rivers, and thus form the majority of large watercourses and the lower reaches of all rivers, total 48 000 km and their productivity is about 200–500 kgkm^{-1}yr^{-1}.

7.3.4 The fragility of the ecosystem and potential threats

Biological activity in a river is determined by ecological factors and is maintained provided that natural or artificial disturbances are not too severe or irreversible, and that they can be tolerated by the species affected.

Damage to the aquatic environment can arise from several natural causes and can cause varying degrees of harm. For example, a large population of

Table 12 Fish production (after Arrignon, 1976)

	Length (km)	Width (m)	Area (ha)	Production $(\text{kg ha}^{-1}\text{yr}^{-1})$	Production $(\text{kg km}^{-1}\text{yr}^{-1})$	Production (t yr^{-1})
Salmonidae watercourses (225 000 km), mainly in mountainous areas*						
Small brooks	150 000	0.5	7 500	30	1.5	225
Brooks	60 000	1	6 000	30	3	180
Rivers	15 000	8	12 000	65	50	780
Cyprinidae–esocidae watercourses (48 000 km), mainly on plains						
Brooks	28 000	1	2 800	80	8	220
Streams	6 000	8	4 800	90	72	430
Floatable rivers	2 500	22	5 600	90	202	505
Navigable rivers	8 765	100	87 650	50	500	4380
Canals	2 320	11	2 550	51	57	130

*Approximately 170 t yr^{-1} of salmon and sea trout should be added to these figures.

carnivorous fish can lead to the disappearance of fry or of other species, but this predatory aspect is only temporary as the predators are then forced to move to a different stretch of river in order to find food. Floods and low-flow periods can cause spawning grounds to disappear and even kill adult fish. Human interference is also important and typical examples are as follows:

(a) Pressure from too many anglers. (In France there is one angler for every 14 inhabitants and in Belgium one for every 60.) This is particularly true for the salmon rivers on which anglers are generally too numerous relative to the productivity. In France, for example, there were 2 869 967 licensed anglers in 1966, although the numbers fell to 2 485 906 in 1976. In practice, the number is probably around 4 000 000 if allowance is made for those who are exempted from licence fees and those who are entitled to fish on private property.

(b) Uncontrolled extraction of sand and gravel from recent alluvial deposits. This can cut the river off from neighbouring marshy areas, thus hindering the movement of the various species into the marshy areas at times of high water and their return at the end of spring. It also destroys the areas where plant life and invertebrates can develop.

(c) Chemical pollution from domestic effluent (e.g. detergent and sewage, which is all too often untreated), industrial effluent from paper and fabric factories, tanneries, etc., and agricultural effluent (e.g. from over-fertilization of the land) can destroy both plant and fish life.

The efforts of environmental protection departments should be recognized in their contribution to improving the quality of water-courses by developing treatment stations and by encouraging industry to save water and to recover harmful effluents from their discharges.

(d) The installation of hydroelectric power stations, including MHPSs, which can cause interference by changing the water-level, by diverting water from the intake dam through the turbine to the outlet, and by the obstacle created by the dam itself.

7.3.5 Impact on the ecosystem

The development of water power has been well documented and maps to the scale of 1 : 86400, begun in 1744 by César-François Casimir de Thury and concluded by his son Jacques Dominique, included watermill sites. By 1840 (cf. Violette, 1902), France had 54444 factories using hydro-power, and Canard (1979) reported a total of 70000 sugar mills, oil mills, paper mills and grain mills as being in operation between 1850 and 1870. He also recorded the decline of water power by giving the total as 50000 in 1890, 20000 in 1900, 8500 in 1936 and 2000 in 1975. In a good stretch of trout-fishing river between Besbré and Loire (Monts de la Madeleine and Bois Noirs), he also recorded 435 sites for 'deriving benefit from the watercourse as the water-level falls', of which only 41 (including five MHPSs) are still in operation.

The repercussions of these mills upon fish life are not at all well known, but they would appear to have had little effect because of the low heads involved and the abundance of the fauna.

Nevertheless, it must not be assumed that MHPSs can be inserted without effect. The desired objective is to provide power in a non-intrusive manner, but they will not achieve this objective if they are built without taking precautionary measures and without taking the environment into considera-tion, without calculating their impact and the damage they will cause, and without regulating for and applying remedies in order to maintain species.

The installation of an MHPS on a river leads to an interruption in the flow of the water. The plant extracts water from the flow at the intake, passes it through the turbine, and later returns it to the river at some downstream point. It can even happen that the flow is not returned to the river at all (see Section 7.3.2). In all cases, however, an MHPS does lead to a reduction in the flow between the intake and the outlet of the tail-race.

This reduction in the flow downstream of the installation reduces both the water flow rate and the area covered by water. The available habitat is reduced and the stretches of running water suitable for the reproduction of salmonidae and cyprinidae are affected. Similarly, the invertebrate popula-tion, which constitutes much of the food supply, will be disturbed and altered even to the extent that fish species have to move from their normal biotope. It is often difficult for them to survive such adjustments.

Depending upon the configuration of the watercourse, e.g. where there are natural obstacles to the flow and the formations of small pools, then under certain climatic conditions (such as a hot, sunny period) too small a flow will cause the rapid growth of the aquatic algae which are harmful to animal life.

The downstream flow of the tail-race can often be reduced to zero when the turbine is not working, and it is then necessary to await the arrival of water from the normal course of the river, where there is only the reserved flow. This type of interruption should be of short duration, say approximately 1 h for a 4 km stretch of river.

In the stretch of watercourse bypassed by the installation, flood-water arriving after a low-flow period strips out the benthic fauna which are not accustomed to such fast water flows. This can sometimes cause virtual sterilization until a new colonization occurs, and it can contribute to environmental deterioration both upstream and downstream of the dam.

The reservoir created above the dam, depending upon its size, can destroy the areas used as spawning grounds. When locks are present, it can happen that at peak demand times, when the income from the national grid is highest, there can be large variations in the water-level relative to the natural water flow and once again the spawning grounds can be affected to a greater or lesser degree. The same situation can occur if the reservoir is flushed too rapidly. Every action which is not well thought out can have repercussions for all aquatic life, whether fish, insects, birds, or plants.

The reduction in the flow rate as the current reaches the reservoir can have several consequences. It can lead to a rise in the temperature of the water which increases the evaporation rate. Also, if the temperature exceeds 18 °C, the water then cannot support salmonidae. It can encourage silting of the river-bed and putrid fermentation, reducing the availability of dissolved oxygen. This leads to a reduction in the river's ability to cleanse itself and to the concentration of any future pollutant which may be discharged into it. It can accelerate the growth of nutritive plankton, with a consequent temporary surge in the number of aquatic fauna.

All these phenomena combine to produce eutrophication which causes profound changes, if not the complete 'biological death' of the water. Bottom-feeders, such as grayling and burbot, gradually decrease in number as do game fish such as perch and pike. Carnivorous fish cannot find food, and the salmonidae, in their turn, are replaced by cyprinidae.

The beneficial effects of introducing these new water-levels should also be mentioned. The biotopes which are created artificially can compensate for the original roosting spots of water or marsh birds which have been displaced by the urbanization or industrialization of their original habitats. They can also sometimes create refuges for fry.

The dam itself can be an obstacle for migratory fish travelling upstream if its height is greater than 2 m. Uncontrolled harnessing of a salmon river with a multiplicity of obstacles can exhaust the salmon so that they are unable to reach their spawning grounds. This has happened in the canalized River Aulne.

It is relatively uncommon for either catadromous or anadromous fish to pass through the turbines provided that grills or bypasses are fitted at the intakes. However, American experience has shown that wounds, ranging from grazes to decapitation, can be caused by the rotating machinery only when the separation of the blades is significantly less than the length of the fish.

Observations on perch and minnows undergoing a pressure change of 1.6 bars have shown, on the other hand, that serious damage, such as internal haemorrhages, eyes popping out, and depression of the swim-bladder, which is sometimes caused by the pressure changes as the fish pass through the turbine. This occurs as a result of defective operation and badly designed draught-pipes and can be traced to cavitation (see Chapter 4). As machinery has improved, these phenomena are now occurring less frequently.

The partial vacuum experienced on passing through the turbine can also cause the release of nitrogen to form bubbles in the vessels and tissues of the fish.

The water which has passed through the turbine, and which has thus been agitated and oxygenated, flows into the river through the tail-race at a speed greater than that of the reserved flow. It thus attracts the migratory fish, and so constitutes a further cause of disturbance to migration.

In conclusion, an objective analysis of the impact should not lead to MHPSs being regarded as calamities as far as fish life is concerned. The degree of impact should always be related to (a) the type of MHPS and the particular design of intake, conduits, turbine, etc.; (b) the size and gradient of the watercourse, the structure of its bed, the water quality, the existing fauna, and, more particularly, the fauna moving through the stretch being considered; (c) the economic context governing the use of the water in the catchment and the added value which can be derived by optimizing the resources and by reconciling the various interests involved.

Countermeasures for the disturbances detailed above do exist, and they should be applied using a sound knowledge of the ecosystem and the legislation developed on the basis of this knowledge.

7.3.6 Countermeasures

The direct or indirect effects of an MHPS installation on aquatic life result from three main causes: (a) the reduction in flow between the intake and the point where the water is returned to the watercourse; (b) the obstacle posed for migrating fish by the intake works; (c) possible eutrophication of the water.

The steps which can be taken to mitigate these dangers include: (a) defining a reserved flow which must be allowed to pass over the works at all times (this is imposed on the project by the regulations); (b) installing efficient fish-passes; (c) maintaining the water-levels.

(a) *Definition of the reserved flow*

Definition of the reserved flow (Chapter 2) is normally based on a give-and-take evaluation requiring a good knowledge of the flows, the biological factors (biocenoses), and the various other uses of the water throughout the drainage area.

Regulations on the reserved flow should be followed and the controlling authority should ensure that the specifications imposed upon the project are adhered to. This latter action assumes that a flow gauge is installed to monitor the flow into the system. The gauge can be either a calibrated sill, or a rectangular or triangular weir which, using calibration graphs, allows rapid calculation of the flow returning to the watercourse immediately below the plant.

(b) *Installing fish-passes*

Fish-passes or fish-ladders should be installed to allow migratory fish to negotiate the intake dams. Examples include the andromodous trout, salmon, shad, eels or any of the cyprinid species, which can travel long distances given the opportunity. Effective operation of the fish-passes is the responsibility of the contracting authority.

A fish-pass must fulfil certain technical and economic conditions. It should allow the movement of every species in the ecosystem by providing a well designed access and flow rates which are suitable for their passage whether by swimming or jumping. Resting places must also be provided along the route. The accumulation of solid deposits should be avoided and efficient maintenance must be provided. An anti-poaching device should be included. The cost should be compatible with the overall cost of the installation and the amount of electricity produced.

Some fish-passes such as Denil fishways require constant effort from the fish, while others, with pools or baffles, require only repeated short bursts of effort with less speed.

The only difficulty which is encountered on the downstream journey arises through high head installations where chutes must be included for the descending fish. For MHPSs, however, fish-ladders can operate in both directions.

Several types of fish-pass can be encountered:

(a) *Oblique and funnel passes*: channels alongside the installation or in separate works 2–3 m above it. If the sheet of water pouring over the dam is too shallow because it is spread over the full width of the dam, it is important to cut a channel for the fish-pass which should not be longer than 20 m for a gradient of 10–20% if it is to be effective (Fig. 51(a), (b)).

(b) *Rustic passes*: When the obstacle to be negotiated is higher, communication can be achieved by installing a separate channel on the river bank. This channel should have a low gradient (<5%) and a pebble bottom to simulate a natural torrent. This can be used in cases where a fish-pass needs to have more space and a faster flow than is provided by the other ladders.

(c) A pass formed of several pools connected by weirs and waterfalls (Fig. 51(c)).

(d) *Deceleration fish-passes* such as that designed by Denil; a channel with a gradient of 10–20% and with the energy being dissipated by blades arranged on the bottom and sides. Because of its low cost, this type of low-obstruction fish-pass is suitable for a dam of height 2–10 m (Fig. 52).

(e) *The Borland fish-lift*: This can be installed economically for dams over 15 m high. It comprises two chambers, one upstream and one downstream, connected by an inclined conduit. The downstream

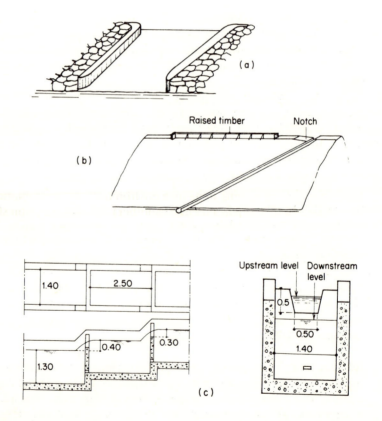

Fig. 51 Various types of fish-pass: (a) funnel pass; (b) oblique pass; (c) a Cipolletti notch-type fish-pass with successive basins, suitable for a flow of $200\,l\,s^{-1}$ and a drop in water-level of 0.45 m (from Larinier, 1977)

sluice-gate is closed periodically and the system is then slowly filled with water. The fish swim up, eventually leaving by the upstream chamber to enter the reservoir. The downstream sluice-gate is then reopened and thus creates a current again.

(c) *Constructing a fish-pass*

Whatever type of fish-pass is selected, the main point is that the access to it has to be highly attractive to the fish.

Fish tend to swim upstream following the strongest current as far as possible until they encounter the dam itself or until they reach the tail-race of the turbine. The entrance to the fish-pass will then be ignored in the face of the more attractive competition.

When siting the entrance to the fish-pass, it is important, therefore, to allow for the routes taken by the fish and the existence of pools which could be used as resting places after unsuccessful attempts. The discharge and, more particularly, the reserved flow at times of low water, will then be directed towards the fish-pass entrance to make it more attractive. Thus, some of the reserved flow can supply the ladder itself, while the remainder can be diverted around the entrance through a channel leading from the foot of the installation and controlled by a sluice-gate.

To summarize, installing a fish-pass requires a prior knowledge of the hydraulic and biological constraints on the watercourse. This includes known data on the following:

(i) The degree to which the fauna liable to be disturbed by the dam is holobiotic (i.e. remaining in its own environment) or amphibiotic (i.e. travelling from one environment to another). This information should include data on the numbers of each species, their routes, their gathering places and their behaviour before the dam was introduced.
(ii) The characteristics of the watercourse, including its pattern, its dependable flow, its morphology, etc.
(iii) The obstacle that is to be introduced; i.e. the type of MHPS to be installed and the materials, height, flood-water spillway, sluice-gates, consumption of energy-generating water, operating periods, amount of variation in the upstream water-level, etc.

For run-of-river MHPSs with heads ranging from 2 to 10 m, most specialists (e.g. Cuinat, Demars, Kiener, Dumont, Larinier, etc.) recommend flat deceleration fish-passes like those designed by Denil and improved upon by Nemeyi, Lachadenède, and Larinier.

The type of fish-pass proposed by Nemeyi and White (1942) is a good example of one which is applicable for gradients between 10 and 20% and discharges ranging from 200 to 800 l s^{-1} (Fig. 52).

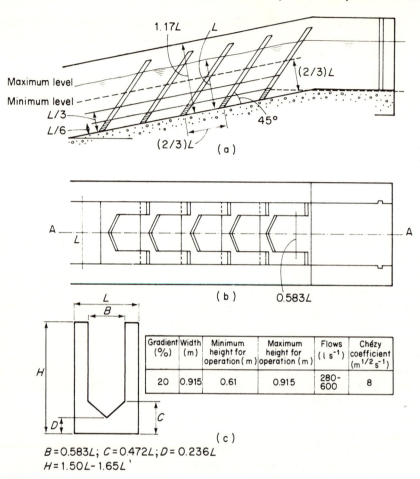

Fig. 52 Characteristics of a Denil fish-pass (from White and Nemenyi): (a) section A–A; (b) plan view; (c) hydraulic characteristics (from Larinier, 1977)

(d)   *Maintenance of the water-body*

As flood-water is not sufficient to cleanse a body of water completely, swamps may form. It must also be admitted that, as maintenance of the plant life along the edges is now reduced, unhealthy species or parasites are not destroyed as they would have been when watermills were in their hey-day.

The maintenance of a body of water implies first of all the maintenance of the surroundings. The plant cover should be partly thinned, allowing for the insolation and the shelter that it provides for insects and for the aquatic fauna. Untreated discharges, discharges of water containing detergents, industrial effluent, and water from filtration plants should be banned or controlled.

The cleaning of the water areas involves uprooting and cutting down plants such as reeds and rushes growing up from the bottom to the surface, or undesirable surface plants, such as cress and water hyacinth. Superfluous sediment, and particularly black muds, should be removed, but the grey muds which encourage the growth of microfauna should be left behind. This operation can be performed using machines of varying sizes depending on the particular case and should be carried out with care. The winter is the best season for such operations and the stretches where salmon spawn should be avoided.

Mechanical cutting can also be carried out in which the vegetation is cut either with a long-handled scythe or with a mechanical mower with multiple shears which is dragged over the bottom, usually in the summer and in any event before the flowering season. Here again, the spawning grounds must be avoided.

Chemical control of the vegetation would take less time than manual methods but should not be employed because of the repercussions it would have on animal and plant life. However, *triazines* which kill plant roots and have a lasting effect are sometimes used.

Some fish such as carp and bream can help reduce the number of plants by eating the roots, and sediment traps can also be built into the dams themselves.

All of the above cleaning operations should be carried out without introducing any new disturbance for the aquatic fauna, which would defeat the object of the exercise. Extraction of the debris resulting from the cutting and mowing operations should be carried out slowly and it should be dumped where it will not cause any environmental disturbance. The refuse should not be dumped downstream in the bypassed section of the river.

(e) *Supplementary provisions*

Among the countermeasures which should be instituted to maintain the fish life and meet the demands of the leisure fisherman, a restocking site should be included. This consists of a widening in the river with conditions suitable for the development of fry and young fish in their first stages of life.

Repopulation of the water in this way may not be sufficient of itself, as fry introduced to a new environment do not always find the conditions that they need to survive and grow.

It would seem to be more effective to raise fry in a fish farm to the size at which they can be taken under angling restrictions, and to release them at the beginning of the fishing season, restocking at the fish farm during this period. This activity creates employment and can be a worthwhile side-product of micro hydro-power stations in tourist regions.

Finally, depending upon the amount of electricity produced, fishing sites can be relocated by agreement with the parties involved.

7.4 CONCLUSIONS

MHPSs can usually be integrated into the natural environment provided that certain precautions are taken and provided that careful consideration is given to locating the various site works.

Conflicts of interest between public or private producers and objectors to the projects, who are in the main the anglers, will be greatly reduced if all parties can be involved in the decisions concerning the details of the installation.

It is important to exert every possible means to ensure the future right of local authorities to produce electricity for themselves and to sell the surplus to the national grid. This will create savings and appreciable income for disadvantaged rural or mountain communities. During an energy crisis, this type of electricity production can be expressed in terms of the tonnes of oil saved.

The rights for recreational activities should also be maintained, principally as they affect anglers and canoeists, by faithfully complying with the specifications which were drawn up after the consultations with all the interested parties.

It should also be stressed that legislation for protecting the natural environment usually requires consultation with the public within the framework of the impact study.

BIBLIOGRAPHY AND FURTHER READING

ANONYMOUS (1978). Les passes à poissons. *La Pêche illustrée*, 1978 (January), special issue no. 550 bis.
ANONYMOUS (1979). Énergie électrique et environnement. Paper presented to the 4ᵉ Colloque Énergie électrique et environnement, Tours, 18–19 October 1979.
ARRIGNON, J. (1976). *Aménagement écologique et piscicole des eaux douces*, Gauthiers-Villars, Paris.
B. G. (1979). La vérité sur les microcentrals. *La Pêche et les Poissons*, 1979 415.
BAUER, J. (1977). Les zones humides artificielles dégradent-elles le paysage ou ont-elles une utilité dans la nature. *Le Courrier de la Nature*, 1977 (48).
CANARD, J. (1979). 500 moulins entre Besbre et Loire. *Cahiers du Musée Forezien*, 1979.
CENTRE TECHNIQUE DU GÉNIE RURAL DES EAUX ET DES FORÊTS (1979). Conception des passes à poissons. Technical information sheet 34 (6), Centre technique du Génie rural des Eaux et des Forêts, Antony.
COURTOT, P. (1979). Le bruit, sa mesure et son utilité dans études d'impacts. *Internal report no. 79/412, Département de Génie géologique*, BRGM, Orleans.
CUINAT, R. (1980). Saurons-nous proteger l'Allier et ses saumons. *Auvergne économique*, 1980 (46).
CUINAT, R. and DEMARS, J. J. (1980). Débit réservé et autres précautions piscicoles imposées lors de l'installation des microcentrals hydro-électriques en Auvergne-Limousin. In *Consultation technique sur la réparation des ressources ichtyologiques*, FAO-CECPI, Vichy.

DELARUE, J. (1980). La pêche dans les Pyrénées. *Revue Adour-Garonne*, 1980 (May).

DENIL, G. (1936–8). La mécanique du poisson de rivière. *Annls Trav. publ. belge*, 1936 (August–October), 1937 (February–December), 1938 (February–April).

DUMONT, B. (1980). Les types d'aménagement hydro-électriques en zone montagnarde et leurs effets sur la vie aquatique. Report of a study instituted by the Centre technique de Génie rural des Eaux et des Forêts, Antony, at the request of the Ministry of the Environment, Paris.

DUMONT, B. (1980). Impacts des microcentrales hydroélectriques sur la vie aquatique en montagne. Paper presented at ALPEXPO, 4e Salon international de la Montagne, Grenoble.

DUROZOY, G. (1979). Microcentrale de Correns-Vallon Sourn (Var). BRGM *notice d'impact*, March 1979.

GUNTEN, G. H. VON (1961) Fish passage through hydraulic turbines. *Proc. Am. Soc. Civil Engrs*, **87** (HY-3), 59–72.

GINOCCHIO, R. (1978). *L'énergie hydraulique*, Eyrolles, Paris.

HURCOMB, L. (1955). *Hydroélectricité et protection de la nature*, Société d'Étude Enseignement superieur, Paris.

LACHADENEDE, B. (1931). Échelles à poissons. *Revue des Eaux et Forêts*, 1931 (9), 763–770.

LACHADENEDE, B. (1958). Les gaves, les saumons, les échelles. *Bull. fr. pisc.*, 1958 (190).

LARINIER, M. (1977). Les passes à poissons. Study no. 16, Centre technicale du Génie rural des Eaux et des Forêts, Antony.

MEVIL, J. (1979). Le scandale des microcentrales. *Nature vivante*, 1979 (18).

NEMEYI, P. and WHITE, C. M. (1942). Report on hydraulic research on fish passes. In *Report of the Committee on Fish Passes*, Institution of Civil Engineers, London.

MUIR, J. F. (1959). Passage of young fish through turbines. *Proc. Am. Soc. civil Engrs*, **85** (PO-1), 23–45.

RAPILLY, J. (1980). Innocentes microcentrales. *La Courrier de la Nature*, 1980 (67).

ROUYER, M. (1975). Entretiens des rivières et aménagement hydrauliques. *Le Courrier de la Nature*, **1975** (39).

VIBERT, R. and LAGLER, K. F. (1961). *Pêches continentales*, Dunod, Paris, pp. 319–337.

VIOLETTE, A. (1902). La disparition du saumon et la question des barrages. *Revue des Eaux et Forêts*, 1902.

WHITE, D. K. and PENNIND, B. J. (1980). Connecticut River fishways: model studies. *Proc. Am. Soc. Civil Engrs*, **106** (HY-7), 1219–1233.

CHAPTER 8

Micro hydroelectric power stations throughout the world

An outline of the role of micro hydroelectric power stations in various countries throughout the world can be formed by referring to recent publications, such as the specialist journals, statistical yearbooks and papers presented at scientific conferences such as that held in 1979 in Katmandu, Nepal, which was organized by the United Nations Organization for Industrial Development.

The position in 20 countries is given in this chapter for the year 1978, under the classification suggested by the World Bank in terms of their gross national product per capita (GNP/C), without prejudging the potential or local skills. The five categories into which the various countries fall cover very wide ranges of GNP/C.

The industrialized countries can meet the demand for electromechanical equipment and some of the developing countries are endeavouring to construct the turbine components while postponing the manufacture of electrical equipment until later.

8.1 INDUSTRIALIZED COUNTRIES, GNP/C > 36 000 FF

West Germany

The total amount of electricity produced was 301 TWh, of which the hydroelectrical contribution was 17 TWh, or 6%. In 1875 the installed power was 118 MW, of which that supplied by MHPSs provided 20 kW, and this doubled every 10 years. By 1910 expansion had stopped, but in 1920 the harnessing of large rivers such as the Rhine, the Danube, the Oder and the Main was begun. By 1940 all the large towns were connected to extensive distribution networks.

By 1945 hydro-power provided 19% of the electricity produced, but has since steadily fallen against fossil fuel stations. Today, there are 500 large power stations in operation under the control of the large utilities and 10 000 MHPSs belonging to independent producers.

The oil crisis stimulated renewed interest in MHPSs, particularly in relatively mountainous areas, where the restoration of abandoned MHPSs is being contemplated and new installations are being initiated under the assumption that profitability is ensured for capacities of over 1000 kW or for capacities between 100 and 1000 kW if the installations are automated.

The cost of an MHPS ($P < 1000$ kW) will be between 10 000 and 20 000 FF per installed kilowatt, inclusive of the civil engineering costs.

No inventory of MHPS sites is yet available, although the importance of this to local industries is recognized.

The best known manufacturers of electrical equipment suitable for MHPSs are Voith and Ossberger. Ossberger hold the patent for the manufacture and export of Mitchell–Banki turbines (Chapter 4). (See Table 13.)

It should also be noted that technical help is given by West Germany to some countries, such as Peru, Burundi and Rwanda, to draw up surveys of hydraulic resources and to install equipment.

United States of America

The harnessable hydraulic potential is estimated as 171 GW and in 1979 the installed power was 61 GW, or 37.5%. The total amount of electricity produced is 2002 TWh, of which hydroelectricity provides 303 TWh, or 15%.

Recently, interest has turned to existing dams which are not equipped with MHPSs, or which are under-equipped. Thus, in its report on the evaluation of the hydroelectric potential of existing dams, the US Corps of Engineers noted that only 4% of the total 49 000 dams higher than 7.5 m were used for producing electricity.

The cost of fitting an MHPS to an existing dam has been estimated to be between 2200 and 8600 FF per kilowatt.

Feasibility studies have been carried out since 1977 by the US Department of Energy on dams with heights of less than 20 m. In 1979, 56 studies of this type were carried out for powers of 110–15 000 kW, covering 31 states.

In 1978 and 1979, New York State initiated the examination of 5300 dams and selected 1670 sites with powers between 50 and 5000 kW and with heads between 4.5 and 13.7 m. The state has provided for the detailed study of 25 sites in the context of the governing bodies, legislation, finance and environmental factors, and has provided for the harnessing of 750 heads in the near future. The project will deal with the standardization of equipment so as to reduce costs, and will try to simplify licensing procedures, which are presently too bureaucratic.

Finally, among the manufacturers of MHPS equipment, particular mention should be made of Allis-Chalmers and James Leffel and Company (now part of the Scandinavian group Nohab-Tampella). (See Table 13.)

France

In 1979 the total amount of electricity produced was about 231 TWh, of which 66 TWh, or 29%, was produced by hydro-power.

In 1977, there were 1060 MHPSs with powers less than 2000 kW, corresponding to a total power of 390 MW and a production of 2 TWh, or 3% of the total electricity produced.

No specific evaluation of the potential of low-power electricity production has been carried out, but it is not unreasonable to expect that at least 9 TWh remains to be harnessed.

The inventory drawn up in 1954 by EDF covered only those sites for large installations with powers greater than 2000 kW.

Since 1977, under the auspices of various government bodies, evaluations of the electricity potential from MHPSs have been carried out using various methods. These surveys have included (a) existing installations and objections to their use (for the Vienne Basin, downstream from Limoges, in the Department of Puy-de-Dôme); (b) existing installations and potential sites (for the central region, Massif Central, Aquitaine and the Argens Basin).

Feasibility studies are under way in the the Burgundy region.

French manufacturers cater for power ranges from 5 to 8000 kW. Of these manufacturers, Neyrpic, C. Dumont, Leroy-Somer, Bouvier and Boussant should be mentioned. (See Table 13.)

Japan

The total installed power in 1978 was 122 349 MW, of which 88 243 MW was fossil fuel, 8007 MW nuclear and 26 099 MW hydro-power. The total amount of electricity produced was 534 TWh, with 425 TWh being thermal, 32 TWh nuclear and 76 TWh hydro-power (which represents 14% of total electricity production).

In September 1978, only seven geothermal power stations were under construction or planned, corresponding to a total power of 218 MW.

MHPSs contribute about 2 GW, if power outputs less than 10 000 kW are included.

Of the Japanese manufacturers, mention should be made of Fuji Electric Co. Ltd. (for powers from 400 to 3500 kW). (See Table 13.)

Sweden

Hydroelectricity (57.7 TWh) represents 72% of total electricity production (78.6 TWh). The harnessable potential has been estimated at 130 TWh.

In 1977, the national industrial agency (SIND) authorized the Swedish association for the development of power stations (VAAST) to begin a survey of existing and potential sites.

There are about 1300 MHPSs connected to the national grid, but they are not all in good working order.

The harnessing of sites with powers ranging from 100 to 1500 kW would supply 500–600 MW, or 2 TWhyr^{-1}, and the power stations already in existence, after some restoration, could contribute a further 1 TWh.

To carry this out, over a period of 20 years, the Swedish government will be investing 6000 million FF to provide 35% of the finance required to modernize or construct the MHPSs.

Between March 1978 and October 1979, 50 million FF were granted to 31 projects, with a total installed power of 13 500 kW. The cost per installed kilowatt varied from 6000 to 21 000 FF.

Agreement on the standardization of equipment has been reached between SIND and the manufacturers KMW and Bofors-Nohab (part of the Tampella-Leffel group). (See Table 13.) Standard units are being developed, e.g. propeller turbines with fixed and adjustable blades.

The obstacle to the further development of MHPSs in Sweden is the legislation on the protection of the environment and the waterways. The legislation currently in force forbids the exploitation of heads which are not already harnessed. Nevertheless, 50% of watercourses with flows ranging from 1 to 20 m^3 s^{-1} have already been allocated to producers of electricity.

Switzerland

Switzerland has made good use of its available hydraulic resources, and, of the total electricity production of 43 TWh, 34 TWh (or 79%) is produced by hydro-power.

Micro hydro-power stations (with powers of less than 8000 kW) produce about 2.5 TWh.

Switzerland provides technical assistance on MHPSs to Nepal.

The companies Charmille SA and Bell-Escher-Wyss should be mentioned as manufacturers of electromechanical equipment. (See Table 13.)

8.2 COUNTRIES WITH A GNP/C OF 18 000–36 000 FF

Finland

The total production of electricity is about 37 TWh, of which 14 TWh, or 38%, is supplied by hydroelectricity.

Finland plans to install MHPSs ($P < 3000$ kW) using Kaplan or Francis turbines, depending upon whether the head is less than or greater than 20 m. Mention should be made of the manufacturer Tampella-Nohab-Leffel. (See Table 13).

A pilot power station of 1200 kW is currently under construction at Karni, being built at a cost of 5000 FF per installed kilowatt.

In Finland a government department has been set up to have particular responsibility for the development of MHPSs, Sahkontuottajien Yhteistyovaltuuskunta (STYU).

8.3   COUNTRIES WITH A GNP/C OF 2700–18 000 FF

Colombia

Colombia has hydraulic resources with a harnessable potential of 70×10^6 kW and the 125 existing large-scale power stations produce 3×10^6 kW of electricity. This hydro-power derives from the basins of the Upper Cauca Magdelena, the Sogamoso, the Sinu and the San Juan, and the target for 1985 is to have an installed power of 7×10^6 kW.

The Colombian government is encouraging the installation of micro hydro-power stations through two organizations, the Instituto Colombiano de Energia Electrica (ICEL) and the Development Centre attached to the University of the Andes (Las Gaviotas).

A plan was drawn up in 1979 by ICEL for the establishment of more micro hydro-power stations, and 35 sites were selected following criteria based on the availability of hydrological data and the existence of plans for the development of agriculture in some river basins. The studies were to be carried out in three stages – preliminary reconnaissance, feasibility and the drawing up of the installation plans. In April 1979, therefore, 27 sites had been deemed worthy of further study. The installed power will be about 60 000 kW, with an average price of 11 000 FF per installed kilowatt.

The Las Gaviotas centre is developing electromechanical equipment, concentrating on Mitchell–Banki micro-turbines and simple propellers.

Spain

Despite climatic hazards and irregularities in the hydrological pattern, as illustrated by flows ranging from 90×10^9 to 34×10^9 m^3 yr^{-1} during the dry season, installed units have a capacity of 13 600 MW with an annual production of 48 TWh.

A study group was set up in 1979 to develop MHPSs in the 250–5000 kW range. An inquiry into existing power stations was undertaken and a survey was carried out to identify those sites which would require only a limited amount of civil engineering and which would not interfere with other installations.

The various hydrographic confederations will also consider the possibilities of placing MHPSs on existing agricultural drainage and irrigation systems.

Italy

Electricity production is 180 TWh, of which 2.5 TWh is geothermal in origin. Hydroelectricity contributes between 50 and 60 TWh, or 30% of the total.

In September 1979, with the help of the EEC, a five-year development plan was instituted in the Mezzogiorno. The aim of the scheme was to benefit from the naturally available resources and to bring needed electricity to the area, which is mainly mountainous, using micro hydro-power stations, windmills, and solar generators.

Nicaragua

Installed hydro-power in Nicaragua totals 107 MW, producing 0.9 TWh of electricity.

Using funding from the United Nations, a programme for MHPSs has been drawn up within the overall framework of an energy programme for Central America.

The first survey covered 60 river basins, and the final selection was based on the availability of data on the hydraulic resources, the energy demands of the area, and socioeconomic considerations.

The basins decided upon for further study were the Rio Negro, Estero Real, Managua, Nicaragua, Punta Gorda, Tuma, Prinzapolka, Coco, Matagalpa and El Sucio.

Panama

In Panama 60% of electricity is produced by hydro-power.

Half the population, i.e. 1 million people, live in rural areas and are dependent upon diesel generators for electricity.

In 1977, a programme to provide MHPSs for communities which were not connected to the national grid was undertaken with aid from the American Development Bank for the construction and equipment of the installations and with funds from the USA for feasibility studies on MHPSs of powers less than 50 kW.

Two MHPSs are being built in central Panama at Coclesito (250 kW) and Santa-Fé (350 kW), and five projects are being considered for installation at Puerto Obaldia (70 kW), Coiba (800 kW), San Miguel (100 kW), Jaque (100 kW) and Rio Sereno (800 kW).

Peru

A survey covering the whole country was carried out recently by the Ministry for Energy and Mines in conjunction with West German consultants. In all, 374 basins are involved, with a potential power of 200 MW, of which the harnessable power is 60 MW.

The aim of the study was to provide electricity to rural areas by means of MHPSs, and to do this it concentrated on heads of 200 m and discharges of $0.2 \text{ m}^3 \text{ s}^{-1}$, or powers of about 300 kW.

The plan for installing MHPSs in the capacity range 50–1000 kW is financed by funding specially designated for electrification by MHPSs, of which 15% is provided from the annual income of ElectroPeru, and the remainder comes

Table 13 List of the principal world manufacturers of turbines for micro hydroelectric power stations

Manufacturer	Address	Type	Unit powers (kW)
France			
Neyrpic	BP 75, 38041 Grenoble (Telex: 320750 F)	All type	1500–8000
		Standard bulb units	150–1500
C. Dumont et Cie	Pont-de-Saint-Uze, 26240 Saint-Vallier-sur-Rhône (Telex: 345501 F)	Kaplan in siphons (heads 2–18 m). all types	100–5000
Leroy-Somer	BP 119, 16004 Angoulême (Telex: 790044 F)	Kaplan 'Hydrolec' (Heads 1–9 m)	5–35
Bouvier	53 rue Pierre-Semard. 38028 Grenoble	All types	500–8000
Boussant	Impasse du Vercors, 38000 Grenoble	All types	500–8000
Switzerland			
Charmilles SA	PO Box 293, 1211 Geneva 13 (Telex: 27334)	All types	60–8000
BFU Division, Escher-Wyss Hydraulique	CH-6010 Kriens/Lucerne (Telex: 78167)	All types	100–2500
West Germany			
Ossberger Turbinenfabrik	BP 425, D-8832 Weissenburg-I-Bay	Banki–Mitchell turbines (heads 1–200 m)	1–1000
Voith GmbH	BP 1940, D-7920 Heidenheim (Telex: 714920)	All types	50–8000

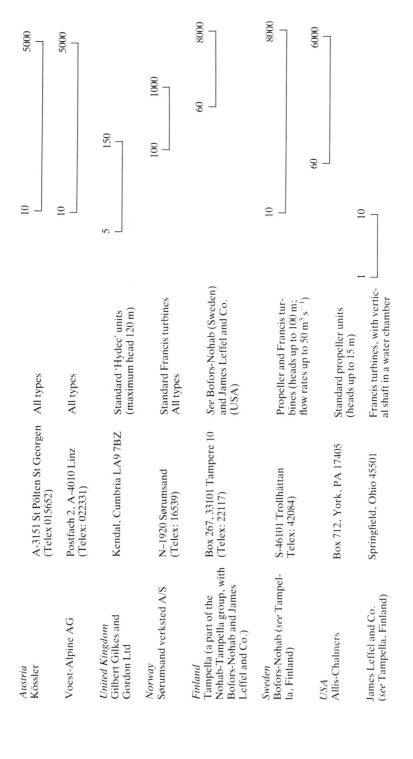

Manufacturer	Address	Description	Range
Austria			
Kössler	A-3151 St Pölten St Georgen (Telex 015652)	All types	10–5000
Voest-Alpine AG	Postfach 2, A-4010 Linz (Telex: 022331)	All types	10–5000
United Kingdom			
Gilbert Gilkes and Gordon Ltd	Kendal, Cumbria LA9 7BZ	Standard 'Hydec' units (maximum head 120 m)	5–150
Norway			
Sørumsand verksted A/S	N–1920 Sørumsand (Telex: 16539)	Standard Francis turbines All types	100–1000
Finland			
Tampella (a part of the Nohab–Tampella group, with Bofors-Nohab and James Leffel and Co.)	Box 267, 33101 Tampere 10 (Telex: 22117)	*See* Bofors-Nohab (Sweden) and James Leffel and Co. (USA)	60–8000
Sweden			
Bofors-Nohab (*see* Tampella, Finland)	S-46101 Trollhättan Telex: 42084	Propeller and Francis turbines (heads up to 100 m; flow rates up to 50 $m^3\ s^{-1}$)	10–8000
USA			
Allis-Chalmers	Box 712, York, PA 17405	Standard propeller units (heads up to 15 m)	60–6000
James Leffel and Co. (*see* Tampella, Finland)	Springfield, Ohio 45501	Francis turbines, with vertical shaft in a water chamber	1–10

Table 13 *(cont'd)*

Manufacturer	Address	Type	Unit powers (kW)
People's Republic of China China National Machinery and Equipment Export Corporation	12 Fu Xing Men Wai Street, Beijing (Cables: Equipex Beijing)	All standard types All types	12 — 500 500 — 8000
	Hong Kong agents: China Resources Company, Bank of China Building, Hong Kong (Cables: CIRIMP Hong Kong)		
	Macao agents: Nam Kwong Trading Co., Nantung Bank Building, Rua da Praia Grande 65A, Macao (Cables: Namkwong Macao)		
Japan Fuji Electric Co. Ltd	New Yurakucho Building, Yurakucho 1 Chome, Chiyo Daku, Tokyo 100 (Telex: 22331 J)	Standard bulb turbine generator units (heads 4–18 m)	400 — 3500

Types of user
Domestic (mainly heating)   1 — 35
Crafts and small industries, hotels, etc.   30 — 150
Independent producers   150 —
Units for restoration of dams   1000 — 5000
National utilities   5000 — 8000

from regional allocations and local resources. The plan is organized by the Ministry for Energy and Mines and by ElectroPeru through its Department for Applied Technology Programmes.

At least 38 sites have been selected since the launching of the programme in December 1978.

The Institute for the Study of Industrial Technology and Technical Standards (ITINTEC) has developed electromechanical equipment specifically adapted to Peruvian requirements.

It should also be mentioned that there are 212 low-power MHPSs already installed, mainly in connection with mining operations.

8.4   COUNTRIES WITH A GNP/C OF 1350–2700 FF

China

China has the largest hydraulic power potential in the world, with a theoretical production of 1320 TWh. Currently, the total annual electricity production is 145 TWh, of which hydro-power provides 67 TWh.

The number of micro hydro-power stations has grown greatly in recent years. Before 1949 there were 26 MHPSs producing 2000 kW (or 76 kW per unit) and by April 1979 this had increased to 88 000 MHPSs producing an estimated total power of 5400 MW, or 60 kW per unit. In one year, 1978, 6000 MHPSs were built, representing a power of 940 MW.

The production of electricity by this means has allowed the growth of agricultural mechanization and has permitted the installation of pumping stations for irrigation and drainage. Above all, it has greatly improved the quality of life, both materially and culturally, for Chinese rural populations.

China is now involved in manufacturing all types of MHPSs, and is beginning to export them through its export agencies in Peking, Hong Kong and Macao. (See Table 13.)

Morocco

Hydroelectric plant of 600 MW capacity produces 1.6 TWh annually, or 30% of the demand. It is planned to increase this production level by 3.2 TWh by installing a further 2000 MW of equipment.

To provide electricity in rural areas, the Moroccan government has also decided to install 10–1000 kW MHPSs (for heads of 10–150 m and discharges of 0.050–5 m^3s^{-1}). The installation cost is estimated to be 10 000 FF per installed kilowatt, and that of the electricity produced to be 0.5 FF per kilowatt-hour compared with the 2 FF per kilowatt-hour of diesel generators. The existing diesel units are being retained, however, as the pattern of water flow in the wadis is so variable.

The Philippines

In 1978, the Philippine government, through the services of the National Electric Power Corporation and the Electricity and Irrigation Departments, carried out a preliminary evaluation of the potential for MHPSs. Of the 4500 potential sites studied, 474 were considered to be suitable for development under satisfactory economic and technical conditions. The Luzon area could support 271 sites, the Bisaye area could support 132 sites, and the Mindanao area could support 71 sites. The installed power would then be 800 MW, or 1.7 MW per unit.

In 1979, financial aid in the form of loans was received to develop the installation of MHPSs, with France providing 200 million FF and the United Kingdom providing 130 million FF.

8.5 POOR COUNTRIES, A GNP/C < 1350 FF

Burundi

Electricity is imported, e.g. from the Mururu power station in Zaire, or produced by oil-fired power stations, whose installed power totals 7000 kW, or derived from hydro-power, with six installations producing less than 11 000 kW.

The consumption of electricity is expanding rapidly but is 95% dependent on imports from neighbouring countries. The government aims to bring electricity to the main towns, bearing in mind that Bujumbura and Gitega are already electrified.

In 1976 the exploitable hydraulic potential was estimated by the United Nations to be 70 MW. Few sites can produce the higher powers, but there are many which are capable of producing 500 kW and are therefore suitable for the electrification of rural areas. Most recent estimates suggest that there could be between 200 and 350 sites for MHPSs with powers less than 5000 kW.

India

In 1979 total electricity production was 22 000 MW.

MHPSs, with powers of less than 5000 kW, are mostly concentrated in the mountainous regions, such as Himalaya, Himachal Pradesh, Sikkim, Arunachal Pradesh, Nagaland, Manipur and Meghalaya. It should be noted that the first micro hydro-power station was installed in Darjeeling in 1897.

There are currently 63 MHPSs in operation, producing a total power of 9600 kW. There are 73 MHPSs under construction, which will provide a further 39 400 kW. It is also planned to install 110 MW of units on the irrigation canals in the Punjab. Thus, there will be 52 sites with heads ranging from 1.2 to 3.4 m equipped with MHPSs of 500–5000 kW capacity, capable of providing local industries with electricity.

India has provided technical assistance to Afghanistan for installations in Bamiyan (750 kW), Samangan (200 kW) and Faizabad (250 kW).

Nepal

A programme for installing MHPSs was established and implemented under the auspices of the State Small Hydel Development Board.

This project was instituted to reduce the amount of wood burnt and thus to preserve trees, so limiting the potentially disastrous erosion of the soil. Local industries could be established, with the development of the artificial fertilizer industry in particular, for the production of nitrates. Improvements in the quality of life were also taken into consideration.

MHPSs are operating at Dhankuta ($P = 240$ kW, $H = 200$ m, two Pelton turbines) and at Surkhet ($P = 345$ kW, $H = 65$ m, three Banki turbines), at a cost of about 9000 FF per installed kilowatt.

In 1979, a technical and socioeconomic survey was carried out for 51 mountain areas covering 32 river basins. The powers envisaged for the MHPSs will range from less than 5 kW (by estimating the potential necessary for supplying 5–10 homes to be 0.6–1.2 kW, at 24 V), 5–100 kW at 230–415 V, and, over 100 kW.

At the moment, studies are being carried out for 10 MHPSs to provide remote centres with electricity. The characteristics envisaged are that $H = 20$–30 m, $P = 200$–260 kW, and the length of the supply channel is 1–3 km. Between 2000 and 4000 people will be supplied, and the cost will be about 6000–23 000 FF per installed kilowatt.

A Nepalese company, Balaju Yantra Shala, is building Banki turbines with a power of about 40 kW for heads of 35 m and discharges of 0.25 $m^3\ s^{-1}$.

Nepal has 170 gauging stations, but torrents are not monitored, so that the hydrological data available is insufficient to find the correct size of turbine.

Pakistan

Hydro-power has been exploited for many years in the mountainous northern and north-west regions and in Kashmir.

Between 1960 and 1973, the state organization – Pakistan Water and Power Development Authority (WAPDA) – initiated the installation of five MHPSs, with a total power of 1130 kW.

In 1973, a construction programme for the north of the country was launched for the installation of 100 MHPSs, and the electromechanical equipment (fifty 50 kW units and fifty 100 kW units) was purchased.

To date, five MHPSs have been installed with a total capacity of 400 kW and five MHPSs are under construction with a capacity power of 700 kW.

WAPDA technicians have estimated that 50 kW of electricity is required to supply a village of 500 inhabitants and that 100 kW is required to supply a village of 500–1000 inhabitants.

The government of Pakistan has authorized the Appropriate Technology Development Organization (ATDO) to begin constructing MHPSs with powers of up to 10 kW. Banki-type turbines are manufactured locally but the generators are imported from China.

Projects involving the installation of MHPSs on canals have been envisaged, where the heads will range from 0.6 to 3 m, even though the maintenance periods will be as long as 1–2 months in each year.

This brief summary has shown the trends and attitudes of some countries classified in terms of their wealth as gross national product per capita (GNP/C).

It is important that the industrialized countries should manufacture standardized equipment and that the products they export are reliable. Overall, it may seem that MHPSs provide a small amount of extra electricity in comparison with the demand, but, at a local level, they have an appreciable added value in disadvantaged rural and mountainous areas. The addition of MHPSs to existing installations is generally taking place or planned for in the near future.

The less wealthy countries, which dream of bringing electricity to their rural areas and of becoming independent of diesel generators, are resorting to MHPSs if their resources permit.

The lack of data on hydraulic resources should not be an obstacle to instituting plans for harnessing watercourses and for their early introduction as methods have been developed for rapidly assessing these resources (Chapter 2).

The various objections raised, on grounds of insufficient available data, or on environmental or administrative grounds, can, in most cases, be overcome. The initial cost of a micro hydro-power station is high, but it is driven by water, a resource which is free, does not need to be imported and is infinitely renewable. Also, various methods of financing MHPS installations are available.

BIBLIOGRAPHY AND FURTHER READING

General works

ATLASECO (1980). *Le nouvelle Observateur*, 1980, special issue.
COTILLON, J. (1978). L'hydroélectricité dans le monde. *La Houille blanche*, 1978 (1–2).
EUROPEAN ECONOMIC COMMISSION (1980). Report of the symposium on the prospects of hydroelectric schemes under the new energy situation and on the related problems, Athens, 5–8 November 1979. Report no. ECE/EP/36.

UNION INTERNATIONALE DES PRODUCTEURS ET DISTRIBUTEURS D'ÉNERGIE ÉLECTRIQUE (1979). L'année 1975, statistiques. *L'Économie électrique*, 1979, special issue.

UNITED NATIONS INDUSTRIAL DEVELOPMENT ORGANIZATION (1979). Seminar workshop on the exchange of experiences and technology transfer on mini hydroelectric generation units, Katmandu, 10–14 September 1979. Issue paper and draft report, UNIDO, Vienna.

Water Power and Dam Construction: papers in various issues since 1976.

France

ANONYMOUS (1977). *Aménagement intégré du Bassin de la Vilaine: Étude des Ressources en eaux superficielles du Bassin de la Vilaine.* Unpublished document no. ETU/09. Agence financière du Bassin Loire-Bretagne, Orléans.

ANONYMOUS (1977). Microcentrales dans le Massif Central; DDA de l'Ardèche, région de Privas. Mise en valeur des ressources en eau de la vallée de la Gluyère; étude de faisabilité. Unpublished report, Compagnie d'Étude de Gestion d'Investissement et de Financement, Coyne et Bellier, Paris.

ANONYMOUS (1977). Microcentrales dans le Massif Central: intérêt humain, aspects économiques, possibilitiés de développement. unpublished document, Compagnie d'Étude de Gestion d'Investissement et de Financement, Coyne et Bellier, Paris.

ANONYMOUS (1979). Les microcentrales dans le département du Puy-de-Dôme. Unpublished document, Département de Puy-de-Dôme, Clermont-Ferrand.

ANONYMOUS (1980). Étude hydraulique de la Vienne dans sa traversée du département de la Haute-Vienne: influence de la gestion des aménagements hydroélectriques sur le régime de la Vienne, en aval de Limoges. Unpublished document prepared by the Bureau Central d'Études pour les Équipements d'Outre-Mer and the Laboratoire d'Hydrologie Mathématique de Montpellier for the Agence de Bassin Loire-Bretagne, Orléans, and the DDE de la Haute-Vienne, Limoges.

ANONYMOUS (1980). Resultats techniques d'exploitation 1979. Report of the Direction de la Production et des Transports, Électricité de France.

BAGNÈRES, M. (1980). Conception des microcentrales. Paper presented at ALPEX-PO, 4ᵉ Salon international de l'Aménagement en Montagne, Grenoble.

BIGNALET, CAZALET, D. (1980). L'énergie hydroélectrique en Aquitaine. Unpublished document, Direction interdépartementale de l'Industrie, Bordeaux.

BOCQUET, P. (1979). Les problèmes d'hydraulique en Corse. 4. Équipement hydroélectrique: l'étude REAM–EDF sur l'inventaire hydraulique de la Corse. *Tech. sci. munic.*, 1979 (12), 603–607.

EECKHOUTTE, M. (1980). Énergie et aménagement du territoire. Paper presented at the Réunion du Groupe de Travail 'Énergie et Aménagement du Territoire', February 1980.

PLAN, M. (1980). Politique énergétique de la région Centre. Paper presented at the Réunion du Groupe de Travail 'Énergie et Aménagement du Territoire', February 1980.

In conclusion

Micro hydroelectric power stations provide an ideal method of producing electricity in rural and mountainous areas where the demand is scattered and relatively low, provided that water resources are available and that the ground has some raised areas.

In the industrialized countries, although micro hydro-power stations will not solve the energy crisis, they do offer the energy savings that governments are encouraging, and their return on investment, which is already very satisfactory, can only increase as the price of oil rises.

To avoid damage to the environment from badly designed installations, micro hydro-power stations should be included in regional planning, which should take account of local objections and the uses to which the water is put, particularly as it affects fishing.

Before detailed plans are drawn up, consideration should be given to the availability of resources, methods of exploitation, matching to the demand and evaluation of the environmental impact. Objective thought has to be given to any conflicts of interest that may arise. In this way, clear and informative data can be drawn up for use in reaching the final decision, which is a political one, on the harnessing of a particular resource.

The addition of micro hydro-power stations to existing dams provided for agricultural purposes or for urban or rural water supplies should be encouraged.

There are undoubted advantages to local authorities in being self-sufficient in the production of electricity and thus reducing the running costs of public buildings and sports complexes in particular, e.g. for ski-lifts and for the heating of sports halls and swimming pools. Furthermore, small industries can be run on this electricity, which, even though it may only be seasonal, is available locally and is renewable, as well as being independent of the vagaries of international politics.

For the industrialized countries, this type of energy represents a source of income and employment, through the manufacture of turbines, electrical equipment, pipework and accessories such as blades and gratings. The developing countries find it difficult to meet the rising cost of oil and the maintenance costs of diesel installations, so that there is a constantly growing need for robust hydroelectric equipment which requires little skilled maintenance and which is thus well suited to bringing electricity to rural areas. The industrialized countries can respond to this need by producing standardized components and supplying much needed expertise.

In some countries, such as Zaire or Colombia, the existence of large hydroelectric projects does not preclude the installation of micro hydropower stations as these avoid the need for connection to the national grid, which could be very expensive and out of proportion with the existing power requirement.

Even those countries which do have oil are trying to benefit from their hydraulic resources, because some communities are too remote from the distribution networks, because the cost of connection to the grid is economically unacceptable or because, ultimately, oil is not a renewable resource and one has to think of the post-oil era.

Capital investment is heavy, but the costs arising from the installation of the turbine can be reduced by using local labour, which thus creates temporary local employment.

Operating costs are minimal as the components are reliable and the water is renewable, albeit with some interruptions which, although predictable, do occur at certain times of the year.

A well regulated growth should be stimulated in this type of hydro-power and every encouragement should be given to those who are striving to exploit the potential of their hydraulic resources.

INDEX